CNC Machining Handbook

About the Author

Alan Overby received a B.S. in Electrical Engineering from Arizona State University. He has always had a hobbyist interest in CNC technology, and has owned, programmed, and operated several CNC routers and engraving machines on a professional level within the signage industry. Mr. Overby was co-owner of Custom CNC, Inc., a company that provided new and replacement controller systems to both individuals and original equipment manufacturers.

CNC Machining Handbook

Building, Programming, and Implementation

Alan Overby

New York Chicago San Francisco
Lisbon London Madrid Mexico City
Milan New Delhi San Juan
Seoul Singapore Sydney Toronto

The McGraw-Hill Companies

Cataloging-in-Publication Data is on file with the Library of Congress.

McGraw-Hill books are available at special quantity discounts to use as premiums and sales promotions, or for use in corporate training programs. To contact a representative please e-mail us at bulksales@mcgraw-hill.com.

CNC Machining Handbook

Copyright © 2011 by The McGraw-Hill Companies, Inc. All rights reserved. Printed in the United States of America. Except as permitted under the United States Copyright Act of 1976, no part of this publication may be reproduced or distributed in any form or by any means, or stored in a database or retrieval system, without the prior written permission of the publisher.

1 2 3 4 5 6 7 8 9 0   QFR/QFR   1 9 8 7 6 5 4 3 2 1 0

ISBN 978-0-07-162301-8
MHID 0-07-162301-9

The pages within this book were printed on acid-free paper.

Sponsoring Editor
Judy Bass

Acquisitions Coordinator
Michael Mulcahy

Editorial Supervisor
David E. Fogarty

Project Manager
Baljinder Kaur, Aptara, Inc.

Copy Editor
Sheila Johnston

Proofreader
Dev Dutt Sharma

Indexer
Aptara, Inc.

Production Supervisor
Pamela A. Pelton

Composition
Aptara, Inc.

Art Director, Cover
Jeff Weeks

Information contained in this work has been obtained by The McGraw-Hill Companies, Inc. ("McGraw-Hill") from sources believed to be reliable. However, neither McGraw-Hill nor its authors guarantee the accuracy or completeness of any information published herein, and neither McGraw-Hill nor its authors shall be responsible for any errors, omissions, or damages arising out of use of this information. This work is published with the understanding that McGraw-Hill and its authors are supplying information but are not attempting to render engineering or other professional services. If such services are required, the assistance of an appropriate professional should be sought.

Contents

Preface ix

Part I The Physical Architecture

1 CNC Machines **3**
   Common CNC Applications 5

2 Guide Systems **23**
   Round Rail 26
   Profile Rail 27
   V-Style Roller 29
   Hybrid Roller Guides 32

3 Transmission Systems **33**
   Screw and Nut 35
   Lead Screw and Nut 38
   Ball Screws 41
   Rotating Nut 42
   Rack and Pinion 42
   Reducers 46
   Timing Belt and Pulleys 47
   Constructing a Pulley-Reduction Unit 49

4 Motors **55**
   Stepper Motors 57
   Servo Motors 61
   Stepper versus Servo: Pros and Cons 63
   Encoders 64

Part II The CNC Controller

5 Controller Hardware **69**
   Enclosure 71
   Breakout Board 72
   Drives 75
   Power Supply 76
   Adjunct Devices for Controller Hardware 78
   Pendant 79
   Wiring 83

vi Contents

6 Control Software 85
    Mach3 Control Software 87
    Enhanced Machine Controller, Version 2 (EMC2) 88
    A Foreword on Computer Operating Systems and Applications ... 89
    G-Code Editors 90
    G Code 90

Part III Application Software

7 The Cartesian Coordinate System 127
    The Table or Mill Topology 130
    Lathe/Rotary Topology 131

8 CAD and Graphics 133
    Raster to Vector Conversion Utilities 137
    Difference between 2D and 3D 138
    Listing of CAD Vendors 138
    Graphics Programs 140

9 CAM Software 141
    Understanding and Using CAM 144
    Generalized Milling Options 151
    CAD and CAM Combination Software 155

Part IV Building or Buying a CNC Machine

10 Choosing a Ready-Made CNC System 159
    Router/Plasma Table 166
    Mills and Lathes 167
    Do-It-Yourself (DIY) 168
    Vendor Listing 168

11 Building Your Own CNC Plasma Table 171

Part V Appendices

A Project Implementation and Examples 187
    Examples of Items that Can Be Produced on a CNC Router 189
    Unlimited Possibilities 218
    Programming Examples 219

B Programming Examples in G Code 225
    Example 1 228
    Example 2 229

C	Engineering Process of Selecting a Ball Screw	231
D	NEMA Motor Mounting Templates	247
	Index	251

Preface

Using CNC, whether on a professional or hobbyist level, is not only an exciting process to be involved in but is also the direction manufacturing is heading. There are a great many facets and stages involved in the end-to-end process of understanding and implementing CNC, and, although there have been several books published on specific aspects or topics (such as G-code programming, building a CNC machine, etc.), there have been no books written that guide the reader through the overall process, that is, until now. It is not the intent of this book to replace any previously written information on this topic nor to delve into any particular area. However, by the time readers finish reading this book, they will have a solid understanding of the entire CNC process from a top-down end-to-end perspective.

More specifically, this book is intended for the following audiences:

- *Academic:* This book will provide the instructor and students a very informative introduction into applied CNC, the various machines, and their uses, along with the necessary tools used in the process.
- *Business owner:* The aspect of moving a small- to medium-sized business, or even a startup company, from a manually concentric manufacturing process into the accuracy and repeatability of what CNC has to offer, can be a daunting task. This book guides business owners in the proper direction to help them understand and decide the ins and outs of automating their manufacturing process. Furthermore, also discussed will be what to look forward to when growing future CNC-based operations.
- *Hobbyist:* There are a great number of individuals interested in the understanding and technical aspects of CNC, but are not exactly sure where to begin—what is absolutely required for the application at hand from both a hardware and software perspective and what is not. There are many free and low-cost software options to choose from that are listed for the reader to appropriately determine what is needed for their particular application.

- *Readers looking for an industry guide:* This book is also intended to be used as a guide, showing the reader that there are certain industry standards within the field of CNC that should be adhered to. There are proprietary hardware and software systems for sale and this book advises the reader as to the pitfalls of using components and systems that are nonstandard. Furthermore, the reader is armed with the appropriate questions to ask the vendors when trying to determine the best approach to take.

Depending on who the reader has previously spoken with or what information they have read, this book will help to augment or clarify what is truly needed for your particular application. This information is to help arm you with the proper information rather than leaving you to rely on what a salesperson is interested in selling you. Often there are low-cost and even free software tools available. These will help you make the determination if certain hardware or software will satisfy your needs, before spending money where you may not need to.

I believe a picture is worth a thousand words. Therefore, I have made every attempt to incorporate illustrations to help the reader visualize what the part looks like and to give an example for reference. Obviously, it would be impossible to include individual pictures of each type of a component, but the main concept is conveyed to the reader with what has been included.

This book also has the following intentions:

- To simplify or demystify CNC for the reader. Where applicable, the intention is to provide the reader with an easy-to-understand, sensible, and logical order of operations.
- To list various hardware and software that I have either previously used with great success or that have been used by companies that have good reputations within the industry.
- To explain in detail the steps and operations used during CAM operations.
- To provide a listing and overview of the commands used in the G-code language.
- To list informative CNC-based Web sites, forums, and additional publications where the reader can obtain more in-depth information on topics covered here.

What I recommend you do as you are reading through this material is to use a highlighter to help you denote the specific items that you find key to understanding the CNC concepts. More importantly, you should keep a steno pad or notebook somewhere close by your computer workstation and CNC machine. Start compiling your own listing of good, known values you

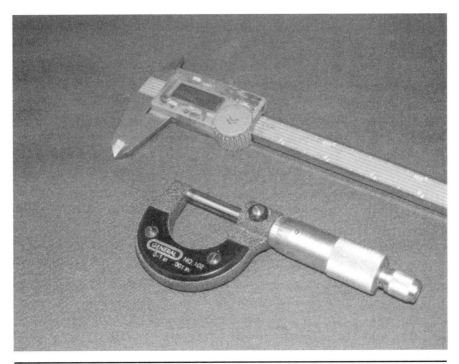

FIGURE P-1 Micrometer and caliper for use in testing the accuracy of both machine and cut parts.

have found for: feed rates, spindle speeds, and cut depths for certain tooling and materials, conventional or climb milling orientations for various material types you encounter, tips and tricks to help you remember various software parameters, etc. It may take you some time to find the optimum cutting parameters for a certain type of material; that is normal. If you have not written down the cutting information, you will have to reinvestigate. An additional suggestion is to make use of an accurate measuring device. Shown in Fig. P-1 are both a micrometer and a caliper. Not only will you need such devices for checking the accuracy of your final part, but they will be invaluable in the initial measurements of materials you are working with (such as thickness). In addition, they will provide accurate measurements for the replication of a given part.

I would like to state that although I will endorse several vendors and their products throughout this book, these are strictly my recommendations. I have no ownership or co-ownership in any of the companies mentioned.

There is one operation of any given CNC machine that cannot be automated, and that is for you to wear the appropriate eye safety glasses! I cannot overstress the importance of wearing protective eyewear. Any of the processes involved in a CNC operation will produce cutting swarf (i.e., dust, wood chips, metal chips, etc). Thus, proper eye protection is a must.

Also keep or install all safety guards on your machinery. Moving and rotating parts can and will pinch and hurt you—the machine will not stop when you yell ouch!

Alan Overby

PART I
The Physical Architecture

CHAPTER 1
CNC Machines

This chapter describes the types of applications that are discussed throughout this text, as well as a list of the most common types of home- and shop-based CNC-controlled applications and their typical construction materials.

Common CNC Applications

This section discusses the various types of applications that can be driven or automated numerically (or numerically controlled by computer). The listing includes the most commonly used applications. Basic and general features found on most commonly used CNC machinery and other applications and their control can be extrapolated from the examples given.

Router/Engraver

Routers come in many sizes and shapes. Depending on what will be produced with a router will have a direct relationship on the proper router head, motors, reduction ratio, speed, gantry height, etc. All too often the term router is generically used to mean various things, but it boils down to a type of machine that uses a rotary process for cutting or engraving. Virtually any sized spindle motor can be used, with its horsepower and rpm capability dependent on the materials and tooling being worked with. Engraving machines can be outfitted with a 1/20 horsepower motor capable of being driven at an rpm of 40,000, whereas a system intended for cutting plywood may have a 40 horsepower spindle with a maximum rpm of 18,000. It is common to find standard woodworking router heads installed on hobby and entry-level machines. This type of motor is quite different, in many ways, by in comparison to a high-frequency spindle head controlled by a variable-frequency drive (VFD). The benefits of using a high-frequency spindle head are many. Among these are the reduced noise of operation, longer life, increased horsepower, and the ability to incorporate an automatic tool changer (ATC).

One of the major differences between these two types of units is their power ratings or the horsepower developed. To help the user understand this difference, the two types of heads are discussed next.

Router versus Spindle Head

A common question from people who are new to CNC routing concerns the difference between a router and a spindle head, as either one of these are generically referred to as a CNC router. Although they both meet the criteria as a router, there are distinct differences between the two. Here we will specifically discuss what each one of these units are and contrast the differences between them.

Router Head

The use of a standard woodworking type of router head is quite common on hobby and entry-level CNC Routers. The reason why this type of motor is used so often is because of its low cost. The type of motor used is referred to as an induction motor. Note that if you spend much time around this type of motor while it is running, you will want to wear some type of hearing protection, as they are quite loud.

These types of router units are intended for general woodworking use and are designed to be used primarily hand-held or inverted in a non-CNC router table. Basically, they are not designed nor intended for use in conjunction with a CNC device. They utilize standard sealed radial ball bearings to support both ends of the shaft and can have rather high amounts of run out. Most have the ability to select the rpm used. The router head shown in Fig. 1-1 has the ability for rpm selection ranging from 10,000 to 21,000 in increments of 2000 and 3000 rpm. They use fixed collet sizes in $1/4$-, $3/8$-, and $1/2$-in increments; reducer adapters are available for smaller diameter tooling (such as, $1/8$ in-diameter bits).

This type of router head usually will claim to having a rather high horsepower – some boasting 3.25 hp, or more. Below, we will discuss both the theoretical and actual wattage and horsepower ratings that can be achieved and conclude with a mention of how the manufacturers derive their claimed values.

Wattage is a product of voltage and current. The theoretical wattage of regular household current (in North America) is:

$$\text{Power (W)} = \text{Voltage (V)} * \text{Current (A)}$$
$$1875 \text{ W} = 125 \text{ V} * 15 \text{ A} \quad \text{theoretical wattage}$$

From the definition that 1 hp is 746 W:

$$1875 \text{ W} * 1 \text{ hp}/746 = 2.5 \text{ hp} \quad \text{theoretical hp}$$

This implies that, theoretically, the most usable power (rated in horsepower) that can be achieved using a standard 15-A wall outlet is 2.5 hp. Note that this value is far short of 3.25 hp.

The value of 2.5 hp is *theoretical*. A typical induction motor will have losses of more than 40 percent. Hence, you might get 60 percent usable power of

FIGURE 1-1
Common woodworking router head.

this theoretical value (which is generous). Reworking our above equation to reflect the typical losses involved yields:

$$125 \text{ V} * 15 \text{ A} * 60\% = 1125 \text{ W} \qquad \text{actual wattage}$$
or
$$1875 \text{ W}/746 \text{ W} = 1.5 \text{ hp} \qquad \text{actual hp}$$

This calculated value of 1.5 hp is less than one-half of the manufacturer's stated horsepower. The reader can be assured that this actual horsepower value is reflective of the most usable power a unit such as this can deliver.

So how did the manufacturer come to their stated value? They are using a measured value of the amperage required at the time of start up for this particular induction motor. This is known as in-rush or start-up current. This occurs for a very brief time as it is a spike in the current and is intrinsic to induction motors. The time the current spikes is so brief that it does not trip the circuit breaker in your electrical access panel. If you use the above equations, you will find that roughly 20 A of current are initially drawn (left to the reader, as an exercise). Nonetheless, in the end, it is nonusable power.

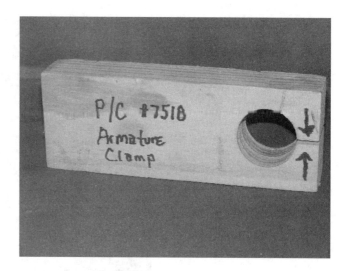

FIGURE 1-2 Bearing removal tool.

If you need to keep start-up costs to a minimum, going with a router head is a viable option. Obviously, depending on several factors, the bearings will often need replacement. This is easily accomplished in-house by the user by making their own simple tool, as the one shown in Fig. 1-2. This tool prevents the rotor from rotating so the unit can be disassembled and the old bearings removed with a bearing puller. These bearings are available either online or at any automotive parts store and will cost $10 to $15 for the pair.

Spindle Head
Spindle heads are physically analogous to a router head, but they work in conjunction with a spindle drive (known as a variable-frequency drive [VFD]) and are frequency controlled to vary the revolutions per minute. Spindle heads are designed and intended for heavier-duty CNC use and typically come with ceramic-style bearings, which are resilient to the higher loads being placed on them. They also yield very low amounts of shaft run out.

Available in a wide range of sizes, they are a constant-torque type of motor that can develop the actual rated horsepower (or kilowatts) as claimed by the manufacturer. Other than the smallest of these units, the power requirements are typically 20 to 30 A at 240 V. Typical sizing for hobby to small-shop production can range from 1.5 up to 7 hp, obviously depending upon the material type and feed rates. The 3 hp unit shown in Fig. 1-3 is made by PDS Colombo and is very popular and reliable. Spindles run very quietly and are available with various options for cooling, including a fan driven from the shaft, an electrically operated fan, and even water cooling (see Fig. 1-4).

It is the function of the variable-frequency drive to supply three-phase power output to the spindle itself. In fact, all spindles are three-phase. It is the power input to the VFD that can either be single- or three-phase. A controller hardware option is to use a spindle-speed controller card to interface

CNC Machines 9

FIGURE 1-3
3 Horsepower spindle head.

FIGURE 1-4
Variable-frequency drive.

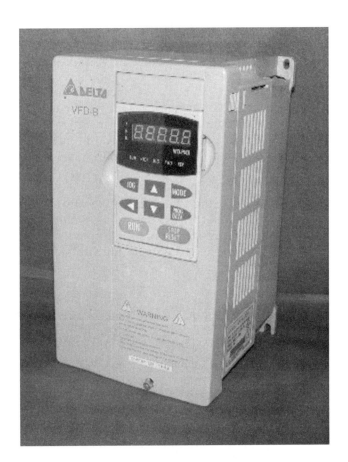

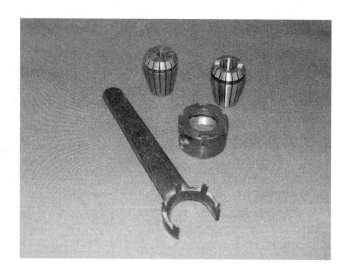

Figure 1-5 ER25 collet system.

with the VFD, allowing the user to turn the spindle on and off, run forward or backward, and to control the frequency or revolutions per minute on a granular selectable level. These controls are accessed either directly by the controller software console or via G-code commands within the cut file. The user can still manually operate the spindle via the interface located on the variable frequency drive itself if the need arises. Some type of spindle interface medium should be considered as standard or a low-cost upgrade option when shopping for a hardware controller. If you are looking to purchase a router table system or upgrade, make sure you have this ability available.

The typical collet system used for these types of devices are ER series (Fig. 1-5), where each compression collet size matches the diameter of the tooling being used. This is quite advantageous as you do not need to purchase tooling that always has common shaft diameters, as with the routers' collet system. There also are some spindles that use a drawbar type of clamping system (pneumatic or electrical), which allows the tooling to be automatically changed out. This is referred to as an automatic tool changer (or ATC). In conjunction with this adaptation on the spindle head, banks of various sized tooling are stored in a fixed location in the router table. The information for each tool, such as diameter, length of cutter, etc., and its exact Cartesian location are stored within a table in the controller software. Via the use of G code, automatic tool change-outs are possible without needing to touch off the Z each time it is accessed.

Machines that are dedicated to an engraving process make use of rather small spindles and servo motors. As the tooling is very small in diameter (generally $1/8$-in diameter or less) the spindles can easily achieve rpms in the 40,000 range. The photograph (Fig. 1-6) depicts an aftermarket engraving spindle that accepts $1/8$-in-diameter conical tooling.

FIGURE 1-6
Engraving spindle.

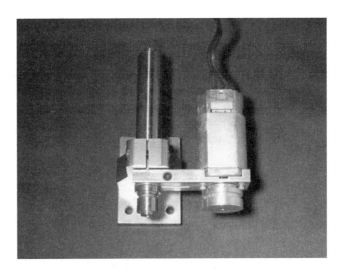

Resolution

For machines with a smaller working envelope, it is not as important to have the ability for high-speed travel. On larger format machines, however, speed plays a critical part in the time it takes for the cutter head to travel from one end of the table to the other. What plays a key role with regard to speed (assuming same motors and drives used) is the amount of reduction being used in the transmission system. For any given generic system, there will be a specific number of steps (think of them as drive signals) that will be associated with producing a certain amount of linear travel – typically 1 in Engraving machines can have typical steps/inch value of 10,000, where a large-format table (8 or 10 ft in length) may have only 2000 steps/inch. The contrast between these two examples has a multiple of five. Hence, for the same rpm motor, the unit with 2000 steps/inch would move five times faster. However, it would have considerably less granularity of cutting ability.

For these types of machines, there will be two choices of transmissions: rack and pinion, and screw. In general, tables that have 4 ft or longer of a working envelope tend to use rack and pinion for the transmission system. When using rack and pinion, some type of reduction unit is required that fits in between the motor and the pinion gear. Without a reducer mechanism (i.e., going direct drive), the resolution of the system will be quite low and the quality of cut will greatly suffer. Short spans and, most often, the Z axis, a lead or ball screw provide the transmission – with lead screws being the dominant choice of the two. Screws are available in various threads/inch values and are generally selected such that a higher resolution value will be given to the Z axis as compared to the X and Y axes. The decision to have higher resolution on the Z axis is typically determined by the application (such as, 3D carving or mold making).

Hold-Down Methods

When performing rotary cutting, the rotating moving bit exerts forces on the material being worked on. To counteract these forces, there are several ways to hold the work material solidly in place. Although any of the below-mentioned hold-down methods will work, each method may not be the optimum solution in each case. It is the user's responsibility to choose and use a hold-down technique that is adequate and safe for each cutting job being performed. Attempting to hold material in place with your hands during a cutting operation is never an option.

Vacuum

Vacuum hold-down is a common method used, in particular, where full-sheet stock is the primary material being worked with. Common industries include woodworking and furniture making while working with full-sheet plywood. The signage industry also uses full sheets of plastics, composites, and thin aluminum sheeting. The vacuum pressure is generated from a unit called a regenerative blower. These blowers are rather large and noisy, and they consume a lot of power during operation. Most spoil boards are plumbed with PVC tubing to create work-area zones. Rather than always creating vacuum across the entire table surface, half- or quarter-sized sheets of material may be worked on using four or more vacuum zones.

T Track Grid Work

For users who typically work with, for example, irregular sized stock, hardwoods, and furniture pieces, great use is made of aluminum T Track, which is embedded into the surface of the spoil board. Various types of fasteners and hold-down clamps are readily available via woodworking sources that are used to securely fasten just about anything to the table. Aluminum is typically used for the track and hardware in the event the cutter bit comes in contact with a hold down.

Double-Sided Tape

Sign shops quite often make use of a sheet of melamine (or similar product) as the surface of the spoil board. Using double-sided masking tape, several rows of tape can be placed on the melamine surface, which holds down sheet stock, such as, aluminum, PVC, and acrylic. This type of tape is rather inexpensive and works as a viable solution to not having a regenerative blower for vacuum hold down.

Fat Mat

Fat Mat is a brand name of self-healing rubberized sheet of material. The material is used mainly in engraving operations where the mat securely holds small pieces of plastic or metal when being worked on. To use, simply place the engraving stock in place and firmly press it down against the mat. Removal of the material typically involves peeling the engraving stock from the mat.

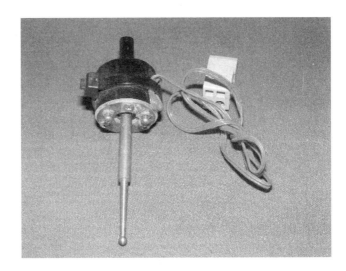

FIGURE 1-7 Device used in 2D probing.

Probing
Devices called probes are used many times as an application to a CNC machine. The machine is instructed to make successive passes over a user-defined area that will be scanned. The scanning process can either be mechanical or optical. The resulting output of the scan is referred to as a point cloud.

Mechanical Probing
Mechanical probes get mounted in a spindles chuck, but at no time ever does the spindle get turned on. The probe (Fig. 1-7) itself has a series of contacts that "make or break" when the probe touches the item it is scanning. The series of the contacts on and off events are concatenated together with specific reference to their X and Y locations. This file can then be replayed and used to reproduce the scanned item.

Optical Probing
There are several flatbed and optical dimensional scanners on the market. Something relatively new, however, is the advent of using a camera mounted on a CNC machine (mill, router, etc.) for very detailed optical scanning. The USB-based microscope camera takes many pictures of the scan item as the axis is moving. Once the scan completes, the software extrapolates a highly detailed mosaic image that represents the scan. This probe, was developed by Tormach and is available on their website: http://www.tormach.com/Product_CNC_Scanner.html. The software works in conjunction with Mach3 controller software.

Rotary A Axis
X-, Y-, and Z-based CNC machines are capable of more than just orthogonal movements. With the addition of a rotary axis (typically designated as the

A axis), horizontal column type of milling/cutting becomes possible. Note that the rotary axis is often referred to as an indexer. An indexer differs from a lathe in that the rotation is not always in one direction and not always at constant revolutions per minute. The CAD and CAM files are laid out such that the file height equals the circumference of the rotary stock. Any changes in *Y* distance of the file equate to a specific number of degrees of rotation. In essence, you end up wrapping the file around the column. The indexer size and column diameter are dictated by the gantry height, if used on a router.

There are a great number of peripheral add-on capabilities a router can have. Many of those discussed are not limited to just a router table, but are generic to CNC tables in general.

Plasma Cutters

Plasma cutters are commonly available metalworking devices that have the capability of through-cutting various types of metals in a single pass of operation. These units come in various thickness-cutting capacities and most often resemble the look and size of a small wire-feed welder. Once the plasma arc is established for the cut, compressed air is used to blow the molten metal through the cut – thus producing the cutting kerf. CNC plasma tables often resemble a CNC router. The notable exceptions in appearance are that the spoil board is replaced with a metal gridwork and the spindle head has an installed plasma torch.

Analogous to establishing a tooling touch off as in a routing or milling operation, an initial pierce height for the material is used to puncture through the metal stock. Once established, an adjunct type of controller device, referred to as a torch height controller (or THC) maintains the proper torch tip distance from the material via constantly sampling the voltage potential between the tip and material being cut. The reasoning behind needing to constantly sample the tip voltage and making subsequent adjustments is that warping of the metal occurs when it is being cut (particularly, thinner materials); not all metal sheet stock lies flat on the table surface and a constant distance between the torch tip and material must be maintained. Furthermore, many sheet metals are not flat initially, but corrugated – hence, another function of the THC is to track irregular-shaped stock.

During normal cutting operation, the motion controller hardware and software have control over both the *X* and *Y* axes for two-dimensional movements, but the THC has control of the *Z* axis for vertical adjustments. The physical interface for the THC type of device is typically via a second parallel port to the computer and controller software. Hence, a total of two db-25 connections are usually required: one for the motion controller and the second for the torch height control. Just as a controller database can store tooling information, various parameters for material type, thickness, feed rates, and plasma-cutting parameters are also typically stored in a database file for easy reference with the plasma operations.

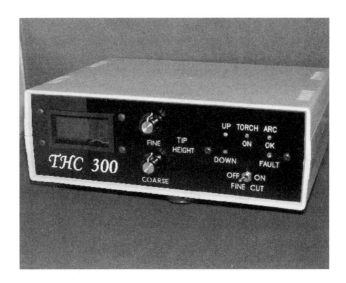

FIGURE 1-8 Torch height controller.

Shops often can be involved in production of materials that can make use of both a router and a plasma table. Invariably the question arises as whether to use one table for both of these types of operations. The recommendation is to avoid using one CNC table for both operations for the following reason: A CNC router table expects the spoil board table to be perfectly flat (or orthogonal) with reference to the Z axis. For use with plasma, there is no spoil board, but rather an open support framework for the molten metal to pass through during the torch operation. Hence, it would be rather difficult and time consuming to dismantle and reinstall a flattened spoil board each time you change out cutting operations. Simply placing a spoil board on top of the plasma grid does not ensure a flat and level surface and would need to be surfaced (i.e., fly cut) each time it is moved.

It is altogether possible to take an existing CNC router and replace its application for use of a plasma torch, which is often done. If you are building or purchasing a table for plasma use, realize that the table's mass and strength do not need to be that of a router's, since little to no force is encountered during a plasma (or laser, water jet, etc.) type of cutting operation in contrast to that of routing or milling. The advanced THC controller shown in Fig. 1-8 is the same one that is used on the plasma table, as shown in Chapter 11 on Building Your Own CNC Plasma Table, and is made by Sound Logic, Inc. (http://soundlogicus.com/). This unit is highly recommended to anyone interested in a new or retrofit for plasma cutting.

Dual Z

There are often times when all cutting jobs on a router will only utilize two cutting bits. In such cases, it may not make sense to purchase an ATC for a spindle, but to add a second spindle or router head. The secondary Z head is easily fabricated and controlled with software, as an independent axis. CAM software can also be configured for post-processor output, designating the primary Z axis as "Z" and the secondary as "A". During the processing of the G-code file, the CNC machine would automatically make use of the proper spindle and cutting tool. This often alleviates the need to change tooling halfway through each file.

Limit and Homing Switches

Switches are not considered an application, but rather a set of controller peripheral devices. The types of switches generally used are micro-switches (Fig. 1-9), which are mounted such as to sense the physical extremes of travel for each axis and in each direction of travel. When using the switches in a capacity of limits, they are intended as safety devices to immediately stop travel of the axis prior to a crash or the gantry running off the end of the table. When looking at the switches with regard to homing, the locations for where each switch trips is a known location. In the event your system experiences a loss of steps condition or the power to your shop goes out, the homing routine will put your machine back into a known working set of coordinates.

The extreme locations where the switches are located and will trip at are known within G code as G53 machine coordinates. The actual working envelope of your spoil board and where typically "$X = 0$ and $Y = 0$" are your G54 table coordinates. When working with switches, you must ensure you know which coordinate system you are referencing.

Micro-switches are single-pole double-throw (SPDT) devices that can be wired as normally closed (N/C) or as normally open (N/O). This topic is covered in more detail later, but make sure you have them wired as N/C devices for safety reasons.

FIGURE 1-9
Micro-switch used for limits and homing.

FIGURE 1-10
Magnetic inductive switch.

One type of switch to avoid, if possible, is the magnetic inductive type (Fig. 1-10). Although they work well as a switch, they do not work well for safety reasons, as they must be wired as N/O devices.

Mills

Mills typically have smaller footprints than most CNC machines and their primary function is working with metals. Mills generally have high transmission reduction ratios, allowing both repeatability and accuracy of 0.0001 in and greater. High-end mills can be found that can perform discrete movements as small as 0.0001 in for intricate work. (Note this is one-thousandth of one-thousandth of an inch!) As mills are intended to work in close tolerances with materials that are highly rigid, the framework of the mill is designed to compensate for high cutting stresses. Mills are exclusively manufactured from cast iron because of its higher mass and incorporate high-lead ball screws for each axis.

Accuracy versus Repeatability

Not specific to mills or metalworking, accuracy and repeatability of any CNC machine are both considered key elements in the design and/or purchasing decision. The accuracy component is a function of the anticipated performance and expectations of the application itself and the repeatability seen as a range or window encompassing it. The tightness of the window it can maintain is the accuracy of the system. The repeatability of the system is determined by its ability to return to the same location time after time. For example, your system can be commanded to move a discrete distance and its accuracy is measured to be off by a certain amount. Accuracy is how close

Lathes

Lathes primarily are intended to work with various metals, although there are CNC woodworking lathes as well. In either case, each has an X and Z Cartesian plane in which work is performed against stock that is rotated by a spindle. Note that in the case of a manually operated metal lathe, the use of back-gears determines the ratio of cutter movement with regard to spindle revolutions. With CNC control, this no longer applies, as the lead screw becomes independent of the spindle. This allows both standard and nonstandard types of threading, as well as the ability to perform wider spans of tapers on the rotating stock.

Construction Materials – Router/Plasma

The overall goal in constructing a CNC router or plasma (i.e., gantry style) machine is typically to have the heavy nonmovable stationary portions to help reduce vibration. Another goal is to have the movable parts be as lightweight as possible (yet strong and stiff enough to handle the intended loads). Thus, faster accelerations will be possible because of lower inertial mass of the movable parts.

Wood Composites

It is common for hobbyists and do-it-yourselfers to employ cheaper and easier-to-work-with materials, such as plywood, plastics, and composite materials [such as medium density fiberboard (MDF) or melamine] to construct their systems. These systems are cheap, fun, and easy to build. However, they have rather short life spans. You cannot compare their metal counterparts and typically they are not consistently accurate.

Aluminum

Another type of common material that can be used is extruded aluminum framing; see http://8020.net/ for an example of this material. There are several companies that manufacture aluminum framing materials, which are available in a wide variety of shapes and sizes. Using this type of material, one can literally build a CNC system much like an erector set. Be advised that these materials and connectors are rather expensive in comparison to an all-steel built and welded counterpart. The benefit is ease of construction for people who do not have the facilities for working with steel. The makers of this type of extruded aluminum can cut to your dimensions and have a large array of fasteners and brackets that can be used for assembly. It is especially advantageous to use the extruded aluminum framing for movable spans, such as on a plasma table or router because of its lighter mass. However, it is important to realize that the expansion and contraction of differing metals can make a difference in the way they are joined together; it is difficult to weld aluminum and steel together. Although this effect is minor, it can affect overall system accuracy as well. In most cases it is customary to bolt together sections of differing materials that cannot be welded.

Joining Materials Together

On the units that use steel and/or aluminum, the type of joining materials can be of paramount importance. For example, a system that is bolted together will, in fact, have a tendency to eventually work its way out of square over time and obviously be less rigid as compared to it being welded together. Of course, there is nothing wrong with drilling and bolting a frame or gantry together first and then following up with stick, wire feed, metal inert gas or gas metal arc welding (MIG or GMAW) and even tungsten inert gas or gas tungsten arc welding (TIG or GTAW) if you have the facilities. If you are constructing a unit from steel and do not have access to a welder, I highly recommend you first square everything up and then have someone weld it for you. There will be a notable difference in the overall rigidity. If you are purchasing a unit (as applicable) you should ask the manufacturer if the unit is a one-piece table or a bolt together. If it is a bolt-together system, either plan on welding it solid or, better yet, consider changing to another vendor that has a welded system.

It is also advisable to have a unit constructed of tubing using any type of material. One material to avoid is c-channel as it tends to be quite flexible, carries a lot of vibration, and maintains loose tolerances with regard to hot-rolled steel tubing.

Tooling

Within the machining and CNC world, the term tooling simply means the cutter you intend to use. There are a great many categories of tooling that can be used and are somewhat specific to the type of machine and material you will be working with. There are so many different types of tooling that are associated with each facet of CNC machining that it would take a separate book to cover them all. What we will do here is cover some of the basic types that are typically associated with wood and plastic routing, as well as engraving. Also bear in mind that the tools listed here are available both as SAE and metric dimensions. The picture shown here shows an array of up-cut end mill bits that range from $\frac{1}{8}$- to $\frac{1}{2}$-in diameter along with some 120- and 150-degree HerSaf v-bit cutters.

- *End mill.* You could consider the end mill (Fig. 1-11) as the workhorse bit most widely used in CNC milling operations. There are types that

20 The Physical Architecture

FIGURE 1-11
Various mill/router tooling.

are specific to the type of material you are working with as well (i.e., wood, composites, solid surface, plastics, and metals). End mills are used in profiling, area clearance, drilling, and inlay types of operations. Among the types of end mills you will typically use: up cut, down cut, compression (up and down cut), roughing, and finishing

- *Ball nose.* These types of bits differ from end mills in that the geometry of the end of the bit is rounded. These bits are also available in various types and are often used in areas such as decorative fluting, engraving, and in 3D finish work.
- *Engraving.* Virtually any style of bit can be used for engraving. However, when you focus your attention more on using a CNC engraving

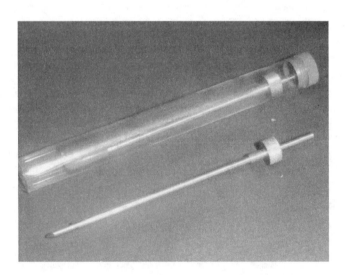

FIGURE 1-12
Conical tooling used in engraving process.

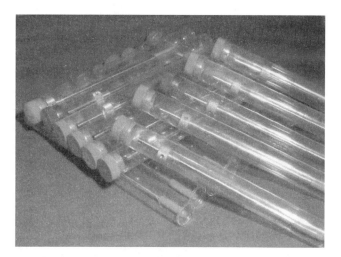

FIGURE 1-13 Set of conical tooling.

machine and working with specific engraving stock (see both http://rowmark.com/ and http://www.inoplas.com/ for just a sampling of material types), then the primary style of engraving tool is the conical bit (Fig. 1-12), although V-style bits are common in engraving operations.

The process of fine engraving uses an entirely different set of tooling. They are aptly named conical, as the tips of the cutters are shaped like a cone, and each have a specific flat spot ground on the tip. These are the long version that work with most engraving machines (this is a full set) and require specialized collets and a high-speed spindle to drive them. They are also available in $\frac{1}{4}$-in shank versions for users who have a router or spindle chuck (see photograph, Fig. 1-13). The particular vendor of these bits, Antares, has a detailed description of the anatomy and intended uses of these bits: http://www.antaresinc.net/FactCutterGeometry.html.

There are also specialized types of these cutters that are used in producing signage that complies with the American Disabilities Act (ADA). Among these are specific angled cutters for material overlay profiling, as well as dot cutters for working with Braille for the visually impaired.

Tooling Systems

For metalworking mills, there is a system that works with common metalworking tooling and is called the Tormach tooling system (TTS; Fig. 1-14). If you do much work on a mill, this tooling addition can improve both your time and accuracy. It works as a series of tool holders that quickly adapt to your existing collet system (R8 or MT3) to change out tools. Each holder fits into the collet the same distance and the protruding amount of the tool is a user-measured distance that gets stored in a file within the controller

22 The Physical Architecture

FIGURE 1-14 The Tormach tooling system.

software. The result is faster tooling changes without needing to perform subsequent touch offs to zero each tool. From personal experience, this is one of the tools you wish you would have invested in long ago. The TTS is available for both manual and CNC-based mills. See the Tormach Web site http://www.tormach.com/Product_TTS2.html.

CHAPTER 2
Guide Systems

In this chapter we will explore the various methods in which the individual axes of the machine derive their motion. All of the systems we will explore are considered to be "linear" guides, as that is the type of motion they yield. Regardless of which type, they all must provide the following:

- Rectilinear (i.e., back and forth) motion along the intended axis
- Smooth straight movement with minimal friction
- Rigid orientation, fixed at 90 degrees (i.e., orthogonal) to the other axes
- Rigid mounting with minimal play between the carriage and guide

There are literally dozens of companies that manufacture linear-motion guides and within each one they will typically carry several different types or styles of guides to fit your needs and pocketbook. Most, if not all, end up being variations on the primary guide styles discussed in this book. For obvious reasons, we cannot look at all of the guides that are commercially available, nor the low-cost types you can build yourself, but we will discuss the ones that are most commonly used on CNC machinery.

Each of the guide systems discussed has certain loading characteristics that vary greatly from design to design. The manufacturers commonly refer to these properties as static and dynamic. Static loads are forces that are always going to be present in the system, whereas forces that come about from variable changes are dynamic. This basic concept is critical in the selection of the guide type and size as it will be the application to the CNC framework that determines the majority of the load capacity. The environment also will determine both the guide type to be used, along with its material composition and lubrication requirements. Where applicable, these will be mentioned with each of the guide styles.

Beyond the basic requirements, it is also advantageous to have the guide rails mounted on the framework in such a way as to not be adversely affected by the cutting swarf that is being produced by the operation of the application itself. Although some debris may make its way to the guides, the goal here is to minimize the interaction that may reduce the life of the guide system or contaminate the lubrication (if applicable). If the installation of the guides results in the collection of too much cutting material, it is customary to use some type of a protective guide along with use of a dust- collection system. In applications where wood or wood-composite materials produce dust, a vacuum-collection system is to be considered imperative, as this type of debris, when airborne, is a known carcinogen. In addition to wearing eye protection, ensure that you protect your lungs by working in a dust-free environment.

Any of the types of guides listed here (properly sized, of course) will suffice to meet the design requirements of most shop-based CNC equipment. However, each one does have its own advantages.

Round Rail

This system makes use of a round rail or rod that provides the linear guide path for one or more bearing blocks to traverse along its length (Fig. 2-1). The diameters of the rail are available anywhere from $1/8$ up to 4 in and are found in lengths up to 20 ft. This type of rail is typically not the same stock as what you would find at your local steel supplier, as it is precision ground to exacting tolerances.

The rail itself can be mounted in either of two ways: end or continuous mounted. Let us first look at the rail or shaft portion of this system with its available mounting options. After that, we will take a look at the bearing and mounting block options that are available. There are two basic types of bearing blocks used with round rail (open and closed). The bearing style will become a key factor when calculating the load capacities for each of the mounting

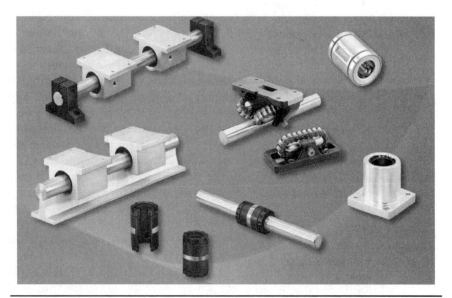

Figure 2-1 Various examples of round rail components from Danaher Motion.

methods. (Refer to Fig. 2-1 for photographs of the various components discussed.) Note that this type of rail system mandates the use of a lubricant, as the bearing blocks have several courses of recirculating ball bearings in them.

End Mounting

In end mounting, only the ends of the rod are supported, leaving the entire midsection of the shafting unsupported. This type of implementation is very low cost and primarily used for short-spanning distances. There are two basic types of mounting blocks used: wall or stand-off supports. The wall support blocks are used in areas where the shafting spans from one parallel surface to another. In situations where there is a rigid substrate running parallel to the shaft and is close by, stand-off supports are used. In either case, a closed bearing style is used.

Often you will find this type of end-mounted shafting used in pairs, with the bearing blocks of each rail rigidly affixed to each other. This greatly increases the overall load capacity, as well as minimizes deflection. The company Danaher Motion (http://www.danahermotion.com/) is one manufacturer of this type of guide system. They have an excellent web-based tutorial with equations to help the user avoid exceeding load requirements.

Continuous Support

The second commonly used option for mounting the round shafting is with the use of a continuous support that has the rail mounted directly to it. It is the continuous support assembly that is affixed to the machine's framework. These supports are generally available as either support-only or as a support–rail assembly. In situations where it is preferred to have one continuous section of rail, multiple stand-alone sections can be butted together end-to-end to provide support. For installations longer than the available length of shaft, end abutment of the support and rail assembly is used. Common lengths for this type of support are generally only up to 48 in. Because of this type of support implementation, the bearing type used is called open. Another excellent source of low-cost round rail components can be found at: http://www.vxb.com/.

Profile Rail

This guide system (Fig. 2-2) derives its name from the shape of the rails' cross section. It has profiled pathways to accept the steel balls contained in the carriage. This is a very popular choice among many users, particularly with designs requiring high-load requirements or precise guide ways. As we will discuss, this system has a great many number of advantages, but it is however, one of the most expensive systems to purchase and can be difficult to install.

Profile rail sizes are determined by their width, measured at the base. Common sizes range from 15 to 35 mm (the sizes are typically always denoted in metric measurement). This guide type is generally known for tight

28 The Physical Architecture

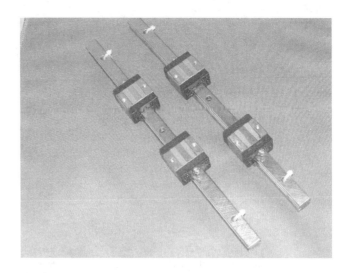

FIGURE 2-2
Examples of profile rails.

tolerances of fit between the carriage and rail, which makes it the best option for many installations. It is, however, this characteristic that can also be a detriment to this guide type's use.

Depending upon the length of rail that needs to be installed, it can be a challenge to getting one run perfectly straight let alone installing a second run parallel to the first within 0.001 to 0.002 in tolerance. Most often, CNC equipment manufacturers use milling equipment to make recessed grooves in the mounting substrate that are ensured to be 100 percent parallel to each other. In addition, the second purpose of the milled areas is to ensure that the rails have the same lateral height. If either of these criteria is not met, the variances in these dimensions can cause the carriage blocks to prematurely wear out.

One advantage of this guide is that its load characteristics are relatively unchanged with regard to the orientation to how the rail is mounted. Coupled with their low-profile space requirements, they often are the best and only design solution.

The carriages (or sometimes called trucks) used with profile rail employ a series of recirculating balls. Hence, they have very low friction and can operate at high speeds.

Carriage options can include lower than standard profile height, as well as availability in wider or narrower widths, which extend their mounting characteristics. Lubrication and swarf removal are typically achieved via optional bolt-on kits that mount to one or both ends of the carriage. There are several variants of the profiled rail discussed above. Among these are a type of system in which the profile of the rail itself very much closely resembles the original type, but it has a specialized type of low-friction coating on its surface. The larger difference will be found in the carriage itself, as some do not make use of steel balls being continuously recirculated. Rather, this carriage's inner profile is also coated with the same low-friction coating material and there is a sliding (versus rolling) action for the carriage to move along the rail. Pacific

Bearing tends to be a leader in this type of technology (http://www.pacific-bearing.com/). HiWin is also a vendor who manufacturers a broad range of good, low-cost profile railing (http://hiwin.com/).

V-Style Roller

This style of linear guide system is a popular choice for use on CNC router and plasma table designs. It is simple in design, implementation, and use and requires little to no maintenance with a long life expectancy. This guide type uses steel wheels with a profiled "V" around the perimeter that ride against a hardened steel track, which has a complementary profile (in reverse) to match. These wheels use a dual row of ball bearings giving them high radial load-carrying ability in a relatively small size.

The reasons behind the popularity of this system are many:

- All parts are available as components – you can easily attach the rail to your own mounting substrate directly on the machine's framework.
- The carriages and railing are available in assembled units in various lengths and sizes.
- Lowest cost of the guide system choices.
- Operates in any environment.
- Low profile.
- Low maintenance.
- Easiest to install.

The wheels and track typically come in four different sizes, where each wheel size in use must match the corresponding track size. Although many implementations are possible, the wheels are mainly used in multiples so that radial load values are a summation of the number in use. This approach easily allows the design to be customized by simple addition of one or more wheels. Worthy of mention is that the wheels have a considerably higher radial load capacity with respect to lateral forces. This implies that proper orientation or direction of the track is such that the greatest loads are radially applied to the guide system.

There are certain applications within industries (e.g., food processing and semiconductor) that tend, for various reasons, to make exclusive use of this type of guide technology. Both the wheels and track are available in stainless steel and, therefore, are corrosion resistant. This works well in areas where cleanliness and periodic wash down of the equipment is preferred. In addition, as the wheels are available as either shielded or sealed, their internal lubrication is not affected. The final advantage is that lubrication between the mating surfaces of the wheel and track is not a requirement of operation so the oil/grease does not end up contaminating a clean room (for example, use in the semiconductor industry).

Figure 2-3 Vector quad showing opposing tracks.

An interesting feature that is intrinsic to this design is its constant wiping or self-cleaning ability. This comes about from the wheel wiping or pushing away any cutting debris as it traverses the track. In addition, felt pads can be mounted on the carriage to provide precleaning and lubrication (if the pads are oiled). This too can extend the life expectancy of the system.

This guide style also has the most ability of customization, as the track can be used as stand-alone or in pairs. In the pairing scenario, they can either face or oppose one another. Figure 2-3 gives an example of the opposing rail situation.

In these situations, lateral adjustments to the wheels are obtained via eccentric bushings. The proper amount of preload can easily be established simply by rotating the hex head on the bushing. (Note: an eccentric bushing has an offset center hole.) The opposing wheels are fixed in place laterally and use a concentric bushing. The concentric bushings have a rounded head and no adjustment is necessary. Figure 2-4 shows examples of both bushing

Figure 2-4 Bushing types: concentric on left; eccentric on right.

FIGURE 2-5
Complete carriage and rail assembly.

types along with some #2 sized V roller wheels. Figure 2-5 shows a complete carriage with rail assembly.

If you are a first-time builder or are designing a CNC system to operate in a dirty environment (routing, plasma, etc.), I recommend using the V-roller guide system. Its lower cost and ease of installation will pay off in the construction process and provide you many hours of maintenance-free operation.

There are several companies who provide both components and assembled systems, but I advise the reader to pay particular note to their pricing. Some vendors producing this style of guide system charge prices that are comparable to that of a profiled rail system – so buyer beware. For reasons of quality, price, and innovativeness, Modern Linear is a preferred vendor of this technology (http://www.modernlinear.com).

A Word of Caution

There are times when individuals or companies use materials, such as angle iron, in lieu of the hardened and ground track for the grooved wheels to ride on. Although this can suffice, there are downsides to this approach. First, angle iron is made of mild steel. Not only will the nonhardened mild steel tend to wear down more quickly, it will wear unevenly. And after some period of use, the angle iron will develop unwanted "hills and valleys" along its length. Second, even though attempts can be made to grind a mating profile along the edge of the mild steel that matches that of the V-roller, it will not be precise enough nor maintain a consistent angle to provide the wiping motion needed to help keep the wheels and rail clean. Last, you will lose the straightness and stiffness in comparison to what the properly installed factory supplied track has to offer. Obviously if you are building your own machine, you have the choice of materials. If, however, you are purchasing a new or used system, please be aware that angle iron is not a comparable substitute for the hardened and ground track that the wheels are intended to ride on.

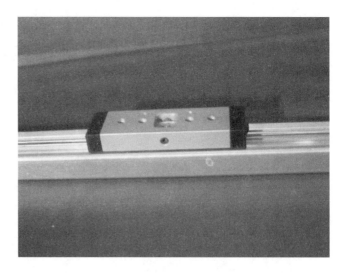

FIGURE 2-6 Hybrid roller guide system.

Hybrid Roller Guides

Working on the same principles as the V-roller technology, some manufacturers use half-moon profile rollers in conjunction with hardened steel rods embedded into a base substrate. The hybrid roller system shown in Fig. 2-6 is made by Pacific Bearing.

This system is only available as fully assembled units and comes in various sizes and lengths. The components are not available individually. You can see from Fig. 2-7 that the center roller is offset and provides the preloading to the bearings.

FIGURE 2-7
Carriage assembly of the hybrid roller guide.

CHAPTER 3
Transmission Systems

There are many ways in which power from a motor can be applied to machinery for various types of movement. Methods such as hydraulic, pneumatic, belt, chain, cable, rack and pinion, and screws are but a few that can be used. However, when dealing with CNC, we are particularly interested in obtaining resolution and accuracy, as well as bidirectional travel. In this chapter, we will discuss the two most common methods in which the rotary power from the motor is translated into linear motion. The function of the transmission system is to translate the rotational power of the motor into linear movement. To meet these criteria, almost all CNC equipment will use either rack and pinion or some type of screw and nut mechanism to produce rectilinear movement. In addition, there can be hybrid systems in which both the guide and transmission are incorporated into one unit.

Screw and Nut

There are several synergies in comparing a lead screw to a ball screw. Among them are how they are mounted (end fixity), nut mounting, and driving method of the screw and its lead (pronounced leed) values (axial movement per shaft rotation)—all of which we will discuss next.

End Fixity

Regardless of the screw style being used, there are four basic methods in which a lead or ball screw can be mounted for use. These are:

- Both ends fixed
- One end fixed with the other end supported
- Both ends supported
- One end fixed with other end free

Before going any further, we need to define the meaning of the terms fixed, free, and supported. The term *fixed* refers to the use of a bearing block that has a set of installed angular contact bearings. Angular contact bearings are a set or pair of back-to-back bearing races that can handle radial loads (such as a radial ball bearing), but also axial loads. Note that standard types of ball bearings (such as radial) handle forces that are radial only—hence the name. When the term *supported* is used, it implies that that end of the shaft has a single radial type of ball bearing supporting it. This is by far the most common type of support found in the majority of hobby and smaller production machines. Obviously the term *free* implies that one end of the shaft is completely unsupported with no mounting.

The type of end fixing required for your installation will determine what type of end milling/turning will be required for each end of the screw. The end types are typically standardized and specific shaft end dimensions can be obtained via the Internet or from any supplier of screws.

The amount of load that the nut can move is greatly determined by the manner in which the screw is fixed, as can be seen in the diagrams. Although there are undoubtedly many applications in which a screw could be used that would carry only a tensile type of force (i.e., force directed away from the mounting block), most CNC implementations utilize cutting forces in both directions (tension and compression). When a screw is under a tension force, the screw will have different dynamic forces associated with it. This is compared to the same situation where a compressive force is applied. During compression, the screw will have a tendency to buckle. At higher rotational speeds, this can induce whip along the length of the screw. *Whip* is defined as the amount of deflection away from an axial straight line that the screw will experience, especially during higher rates of revolution. Commonly referred to as *critical speed*, it is the highest revolution rate of the screw that can be obtained before whip occurs.

Lead Values

Lead (pronounced leed) values are simply the amount of axial travel that the nut will experience during one single rotation of the screw shaft. When shopping for a screw, you will find that there are a great many different lead values available for any given diameter of shaft. For example, if using a microstepped motor (with 2000 steps per motor shaft rotation) directly coupled to a screw with a lead value of 0.2 in, your system would be required to produce 10,000 steps for each inch of travel. Note that an application that requires this many steps per unit of travel will not move or cut very fast (maybe 150 to 200 in/min or ipm) so it would not function well in a large-format table router design. It would, however, perform quite well as an engraving machine where higher resolution is required. If a screw with a lead value of 0.5 were chosen, then two motor shaft rotations will be required to move the nut 1 in of travel. Hence, the system's motor tuning value would be 4000 steps/in. Using the same controller system and drives, this system would be able to achieve travel rates around 600 ipm. Note that 600 ipm equates to 10 in/s (ips) (i.e., 600 ipm = 600 in divided by 60 s = 10 ips). For a large-format table router (say 4 by

8 ft cutting envelope), this is a desirable feed rate, as longer distances of travel can take a lot of extra time to move.

There are a few methods in the determination of both the shaft diameter and lead value needed for your application: borrow from someone else's design, seek advice from a supplier, or calculate your own. If you choose to borrow from someone else's specifications and design, make sure it will be used in the same capacity (speed, shaft size, end fixity, etc.) as your application calls for and that its performance is satisfactory.

Other items of interest to some will be cumulative lead errors that can occur over the travel length of the screw you select, as well as life expectancy.

Nut Mounting

Whether you are using a lead screw nut or a ball screw nut, there needs to be some method of attaching the nut to the movable portion of the axis. Both types of nuts generally use a threaded portion of the nut or incorporate a mounting flange to attach to. Your specific implementation will determine the size and type of bracket to go in between the nut flange and axis carriage. Custom brackets are generally user built from some type of angle stock (aluminum or steel).

Driving Method

In most instances where screws are used, they will be direct driven from the motor to the screw shaft via a coupler. These types of couplers can come in several design types broken down into either rigid or flexible. The rigid-style clamp used between the motor shaft and the screw shaft form a rigid coupling bond via use of a one-piece clamp-on mechanism. The flexible type of clamp also has separate coupling hubs that get affixed to each shaft end, but have a urethane spider gear meshed in between the two coupling hubs. What this allows the user to do is more easily remove the motor from the shaft without having to unscrew one of the coupling ends, as is the case in the rigid type. Of the two types, the rigid coupling has a much higher load-carrying capacity, but is considerably more expensive because of the additional machining required to produce it. There is also an alternate indirect method. The drive lead and ball screws utilize timing pulleys mounted to the screw shaft and motor drive shaft with a properly adjusted timing belt between the two pulleys. This type of arrangement can have the advantage of remotely mounting the motor to another location and/or the ability of changing the overall drive ratio between the motor and the screw. This concept is further discussed below in the section entitled "Timing Belt and Pulleys."

Screw Lead

Another synergy between screw types is the lead or turn value and how it's calculated. Depending on the screw supplier and their naming convention,

38 The Physical Architecture

Lead	Turns/inch	Steps/inch
0.1	10	20,000
0.125	8	16,000
0.2	5	10,000
0.25	4	8000
0.500	2	4000

TABLE 3-1 Summary of Turns/Inch and Steps/Inch of Common Lead Values

they will denote the axial travel of the nut as either lead (usually given in decimal or inch fraction) or in turns (per inch of travel). Note that each of these is simply the reciprocal of each other.

Table 3-1 above summarizes the turns/inch as well as the steps/inch (10 × microstepping assumed) of some of the most common values use. As you can see from Table 3-1, most builders for a general-purpose machine should opt for a screw with a 0.500 lead value. Others, who are interested in higher resolution, such as for an engraver or milling machine (which don't need the higher traverse speeds), would tend to opt more for a 0.250 or 0.200 lead on their screw. Obviously a timing belt with pulleys can be used to further fine tune your application. Note that calculating turns and steps/inch for ball screws is performed the same way.

Lead Screw and Nut

Lead screws have a thread form known as Acme. These are commonly referred to as Acme screws. Depending on the supplier of the screw (i.e., lead screw specialty supplier versus a machine shop parts supplier), there are some notable differences—other than price. Although they both have Acme thread styles, the screws that come from a general-machining supply company will typically use a lower grade of steel, and the actual thread form will not adhere to as close of tolerance as compared to those found from a lead screw supplier. Companies that specialize in lead screws will often offer screws that have high-wear, low-friction surface treatments applied to them.

Lead screw nuts will either come as a one- or two-piece design. The one-piece style is very much like a regular nut, but has a longer body with more threads engaged on the screw at any given time. Figure 3-1 shows a lead screw assembly with a one-piece nut.

These nuts are available in either bronze or, more commonly, some type of low-friction plastic, such as Acetal. The second and more commonly used type of nut for lead screws comes in a two-piece design and are antibacklash in nature (see Fig. 3-2).

Although it is not always apparent that these are actually of a two-piece design, sometimes, but not always, the manufacturer will enclose the nut with an outer covering. *Backlash* is the amount of play or movement the nut

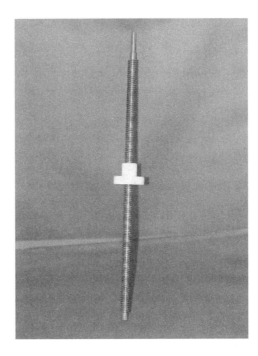

FIGURE 3-1 Lead screw assembly with one-piece nut.

experiences when the nut is changing direction axially along the screw. The two-piece design of an antibacklash nut (Fig. 3-3) actually uses two nuts that are fixed to one another from a rotational perspective, but are separated from each other via spring force. What this achieves is a force from one of the nuts against the screw threads always in a forward direction. At the same time, a backward force is exerted against the back side of the screw's threads. The

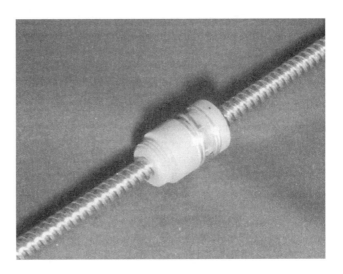

FIGURE 3-2 Lead screw assembly with two-piece antibacklash nut.

FIGURE 3-3
Close-up of an antibacklash nut.

result is very little to no movement or backlash between the nut and screw. Typical backlash values for lead screws and nuts range from 0.010 down to 0.003 in/ft.

Another characteristic that is inherent in using lead screws is a result of the friction generated between the nut and the screw. Lead screw nuts have a rather large ratio of contact area between the nut and screw. Even though efforts are made to use slippery materials to make the nut, as well as low-friction coatings applied to the lead screws, these larger contact surfaces equate to a resistance in moving the nut while the screw is being rotated. When shopping for lead screws and nuts you will see this referred to as *efficiency*. Also worthy of noting is that for any given lead screw diameter and load-carrying capacity, the percentage of efficiency is directly proportional to the amount of lead the screw has. (Recall that the lead of a screw is the amount the nuts travel based on one rotation of the screw.) Lead screws with a low lead amount will invariably have threads that are much closer together and will end up having a lot more surface area in contact with the nut and screw at any given time. Contrast this to the same diameter screw with a larger lead value – you should be able to envision that the screw threads will be much farther apart from each other. The end result is lower friction and an increased efficiency percentage.

This resulting efficiency can be good or bad. You can use it to your advantage, for example, in cases such as the Z axis on a CNC router. With an efficiency of even 60 or 70 percent, this will negate the need for your vertical axis system to have a counterbalance weight or spring. Some type of spring or other force is required in a vertical application so that when power to the axis is removed, the actual spindle or router head does not slide down to the bottom of its travel. This is why it is quite common to see a table router with one type of transmission on its X and Y axes, and a lead screw installed on the Z or vertical axis.

There are a great number of first-time CNC machine builders who use common thread lead screws and end up with poor results in the area of speed. The reason is because standard lead screws typically have a higher number of turns/inch. For example, if you use a ⅝-in-8 screw, with a root diameter of ⅝ in it will travel 1 in per eight revolutions. Therefore, in 1 in of travel, your drivers need to produce (assuming microstepping) 8 × 2000 or 16,000 steps/in. Obviously this value will be dependent on whether you are building an engraver or a gantry-style router, but this number is way too high to achieve travel speeds much over 2 ips. It is a rather straightforward and simple fix—using the inverse of a reducer, a ratio increaser. (See the "Timing Belt and Pulleys" section later in this chapter for more information.) By using a 3:1 pulley increaser, you can use the same motor and lead screw example and have 5,333.333 steps/in. This step value will produce much faster cut and rapid speeds, while still having great resolution.

Rather than purchasing a common teeth per inch (TPI) Acme screw, you could also purchase one that already has the proper number lead for your application. That way you can design both intended speed and resolution into your machine. For example, in looking at the lead screw availabilities listed on the http://www.thomsonlinear.com/ website, you will find that ⅝-in lead screws are available in the following lead values: 0.100, 0.125, 0.200, 0.250, and 0.500.

Ball Screws

There are several advantages that ball screws have over lead screws. Among these are efficiency, precision, higher-load capacity, longer life expectancy, higher operating speeds, and ability to back-drive via a rotating nut.

Ball screws are the higher cost and higher performing of the two choices of screws. Unlike their counterpart, ball screws all have a very high percentage ability of transferring almost all of the rotary power into linear motion. In fact, ball screws typically have efficiency ratios at 90 percent or greater, whereas lead screws rarely achieve even 75, with 50 percent being more common. This is primarily achieved from the method in which the ball nut interfaces with the screw ball bearings. Rather than having a lot of material from a lead screw nut friction sliding against the screw, the ball bearings rotate or spin within the tooth form of the ball screw, greatly minimizing the force required to rotate the screw. These ball bearings are all fit within a channel (or channels) that reside in the ball nut. Hence, not all of the bearings are in contact with the thread form at any given time, but recirculate throughout its circuit.

The precision of a standard ball screw will be higher than that of a precision-class lead screw. Standard values of precision are in the 0.004-in/ft range, with higher precision classes available that reach 0.0005 and even 0.0002 in/ft. Coupled with having tighter tolerances of error per unit length, ball nuts have considerably less backlash, as their inherent design causes a constant force to be applied radially inward forcing the rotating steel balls into the thread form of the screw, thus taking up any backlash.

Furthermore, this style of nut design yields greater load-carrying capacities, life expectancies, and achievable travel rates. Most manufacturers supply various graphs of anticipated life expectancy of the screw and nut assembly versus load, maximum travel rate versus the length of the screw and the engineered compressive loading versus length, both of which are also dependent upon the type of end fixity used.

Another feature you will find common to ball screws is the number of starts or independent thread forms on the screw. Typical start values are one, two, and four. As the lead value of a screw is determined by the product of the screw's pitch times the number of starts, having an increased number of starts has the effect of doubling or quadrupling the lead value. Thus, be aware of this during your selection.

Rotating Nut

There is one exception to the driving method as discussed earlier and that is to drive the rotating nut while the screw is prevented from turning. This is becoming a rather popular option among users who employ long lengths of screws on their machine (typically 6 ft or longer). The advantage here is that whip is minimized. The nut itself will have a timing pulley attached and is driven via an under-table gantry-mounted motor via a timing belt.

Rack and Pinion

The use of a rack and pinion is generally the most popular method of mechanical translation methods when longer distances are required, as is the case in large-format routers and plasma machines. This is primarily because of cost issues, as rack and pinion is undoubtedly more cost effective than using either lead or ball screws.

There are two methods in which to use the rack and pinion combination: The most common method is to affix the rack to a stationary portion of the machine frame and drive the pinion, which is fixed to the moving part of the machine. The second involves fixing the pinion gear in one location and having it drive the gear rack. In an example of the latter implementation the bed of the machine moves back and forth, instead of having a movable gantry (Fig. 3-4).

There are two offerings of rack and pinion that are commonly used, which are based upon the pressure angle (or PA) of the pinion and rack gear-tooth profile. The most commonly used of the two is the 20-degree PA, as it has a higher load-transfer capability, as compared to a $14\frac{1}{2}$-degree PA. Regardless of which pressure angle design is used, each will have a range of pitch available to choose from. *Pitch* is defined as the distance from tooth to tooth

FIGURE 3-4 Section of rack with two pinion gears.

(or valley to valley) of the gear form. Note that: because of the differences in the pressure angles, it is not possible to interchange or mix the two. For example, you cannot use a 20-degree rack with a pinion that has a 14½-degree pressure angle. They are like oil and water—they just do not mix. To simplify matters, manufacturers have adopted a standard for each pitch to equate to a corresponding face width of both the pinion and rack. For example, a rack that is 20 PA and 20 pitch will have a ½-in face width, compared to a 14½ PA 20 pitch of ⅜-in face width.

If you are designing a machine and you decide you want 20-degree PA with 20 pitch, the only unknowns left to deal with are the number of teeth of the pinion gear and the diameter of the borehole to rotate the pinion gear. Deciding on which pinion size (i.e., diameter) has a direct relationship on the number of steps/in of travel that your machine will have. The typical number of pinion tooth choices ranges from 15 up to 40 teeth. These values equate to gear diameters of 0.75 to 2 in. The values you will be working with are not the overall diameter of the pinion gear, but rather the pitch diameter (PD). The *pitch diameter* of a pinion gear is defined as the number of teeth in the gear divided by the pitch (which for our example is 20 pitch). Note that the actual PD lies one-half a tooth *within* the actual pinion diameter. As we are interested in obtaining the number of steps per unit of travel (i.e., inch), we simply multiply the PD times pi (i.e., 3.1416). This will yield a value of linear travel (in inches) per single revolution of the pinion gear. For this example, we will assume that a standard microstepping driver is being used (10-step multiplier), which gives 2000 steps per motor revolution. Now if you take the value of travel per revolution and divide it by 2000, you are given what is in/step. As we are interested in steps/in, we simply take the reciprocal of this value to get the calculated steps/in. An example will show that this is not really as complicated as it sounds (plus, as discussed earlier, the math would be simple!).

Further suppose you are interested in designing a CNC router that will operate in the 400- to 500-ipm cutting range and have rapids of 600 ipm or higher. In this example, you are shooting for something in the range of 2000 to 4000 steps/in (using microstepping). For this machine, we have decided on using 20-degree PA with 20 pitch ($\frac{1}{2}$-in face width on the pinion and rack).

First, we calculate our PD:

$$PD = \text{number of pinion teeth/pitch}$$
$$= 20/20 = 1$$

To get the inches of travel per revolution of the pinion gear:

$$\text{in/rev} = PD \times Pi$$
$$1 \text{ in} \times Pi = 3.1416 \text{ in}$$

To get in/(micro) step:

$$\text{in/step} = \text{in/rev}/2000$$
$$3.1416/2000 = 0.0016$$

Taking the reciprocal yields steps/in:

$$\text{steps/in} = 1/\text{in/step}$$
$$1/0.0016 = 636.6198$$

Did we make a calculation error? We clearly fell short of our goal of 2000 to 4000 steps/in. If we used a larger pinion, the steps/in value gets even lower! In actuality all of our calculations are correct. In fact, this is what can be expected if you have a motor *directly* coupled to the pinion. In this example, the carriage of the gantry will move rather quickly, but the resolution will suffer terribly.

Note that there are individuals and even companies that sell CNC equipment that have direct-drive rack and pinion setups such as this. A system such as this will have very choppy looking curves. If you are in the market for a unit – beware!

The resolution or fix to this problem is to use a reducer. A reducer is a mechanical device that reduces the rotational shaft output relative to the input. Common reduction ratios for use with rack and pinion gearing are in the range from 3:1 to 5:1. What this implies is for every (example) three turns of the input shaft, the output shaft will rotate once. Now if we applied the 3:1 reduction to the example above, we would get three times the 636.6198 or 1909.8594 steps/in. Now, that's more like it! And if we used a 5:1 reduction, we would end up getting 3183.099 steps/in. Obviously, we could use a 5:1 reducer along with a larger pinion as well.

Having this information, let's modify our formulas to include the use of a reduction unit (4:1) along with an example using a 30-tooth pinion gear:

First we calculate our PD:

$$PD = \text{number of pinion teeth/pitch}$$
$$30/20 = 1.5$$

To get the inches of travel per revolution of the pinion gear:

$$\text{in/rev} = PD \times \text{pi} \quad 1.5 \times \text{pi} = 4.7124$$

To get in/(micro) step:

$$\text{in/step} = \text{in/rev}/(\text{reduction} \times 2000)$$
$$4.7124/(4 \times 2000) = 0.00059^*$$

Taking the reciprocal yields steps/in:

$$\text{steps/in} = 1/\text{in/step}$$
$$1/0.00059 = 1697.6488$$

At this point, we can either increase the reduction ratio or recalculate using a smaller pinion gear.

Rack and pinion transmission systems have a very high load-carrying capacity for transferring rotary into linear motion. In addition, they are engineered such that the pinion gear is made from a softer material than that of the rack. Hence, the consumable that wears out over time with use is the pinion gear. Obviously with any geared system, the user must maintain lubrication to minimize wear. Worth noting, but of minimal importance, is that a smaller diameter pinion will spend more time engaged in the rack for a given distance compared to a larger pinion. That means a smaller pinion will tend to wear more quickly. However, the design intent of this type of a system is that the pinion is intended as a user consumable. The number of pinion teeth engaged in the rack is not considered significant since pinion sizes are only manufactured down to a certain number of teeth and the smaller gears are still more than capable of maintaining a high level of load capacity.

Anytime there is a gear-to-gear or gear-to-rack type of engagement, there will be some amount of backlash involved. As long as you have lubricated the rack on a regular basis (depending on use and dust, etc.), you can expect very low backlash. There are times, however, when even this small amount of gear lashing is unacceptable. Setups are available where parallel racks with

---

*A calculator was used; and this is a rounded value.

side-by-side pinions reduce the backlash to less than 0.001 in. Some of these systems continue to use straight-cut spur gears, while others have gone to a helical cut, yet others employ a split-pinion for use on the same gear rack. Regardless of the approach or manufacturer, obtaining zero backlash in a rack and pinion system is achievable, but it will cost you a fair amount of money to implement.

Furthermore, I will add that there are many varying materials that both the rack and pinions are made of. Industries regulated by the Food and Drug Administration (FDA) and United States Department of Agriculture (USDA) use this type of setup for washdown areas, where nylon and stainless steel are the materials most commonly used. Steel is the least expensive and most common material type used on most shop CNC equipment.

Reducers

There are two common methods in which to alter the input ratio of a screw or pinion: geared or belt driven. Each of these have associated pros and cons to deal with, and you will quickly find that cost plays a major deciding factor in which way to go. We will discuss the geared type first; we will finish with timing pulleys.

Geared Reducers

Within the geared type of reducers, you will most commonly find two types: planetary and offset spur. Gear reducers fit into a nice compact-sized housing that usually has the same footprint as the motor frame size (i.e., NEMA 23 or NEMA 34). The two units get bolted together with the output of the motor driving the input gear of the reduction unit.

Planetary Gear Reduction

These types of units are very commonly used for a number of reasons. They are compact, readily available as a bolt-on device, have very high load-carrying capacity, and are relatively quiet during operation. In shopping around for this type of reducer you will find a quite diverse range of pricing. For the most part, the cost differential will be reflective of the degree of backlash. There are some commercial applications where both high cutting forces and precision tolerance is a must. In a situation such as this, the use of planetary gear box warrants itself and offsets the approximate $500 per unit cost. These are also available in a wide range of reduction capacities going from 2:1 up to 100:1.

Offset Spur Gear Reduction

This type of reduction unit derives its name from the output shaft being *offset* from that of the input shaft (Fig. 3-5). Within the housing are a series of straight-cut spur gears much like those you would find in an automotive standard transmission. (The spur gears look like pinion gears.) The rated

FIGURE 3-5 Offset spur reducer coupled to stepper motor. Note the *offset* output shaft.

load-carrying capacity of these units is generally high enough to meet the needs of most CNC machines (with exception of a metal milling machine). The drawback to these units is that they inherently have large amounts of backlash with no ability to adjust the gear-meshing preload amount.

There are a few companies that offer this type of reducer, however they are typically only available as a motor plus reducer combination unit. If you are in the market to purchase a CNC machine (or parts for building your own) and are interested in keeping backlash to a minimum, then these offset reduction units are to be avoided. Like the planetary-style gear boxes, these too are rather compact to work with. Unlike their planetary counterpart, these are very cheap. You can easily find a motor/offset gear reducer combination at a price less than one-half that of just one planetary reducer. Be cautious; the old adage of "you get what you pay for" applies here.

Timing Belt and Pulleys

This type of reduction derives its name from the type of components it uses: timing pulleys and a timing belt (Fig. 3-6). These types of components are available in various levels of pitches. Recall that pitch is defined as the distance

48 The Physical Architecture

FIGURE 3-6 Timing belt and pulleys.

from tooth to tooth. For a given pitch, there are a range of pulleys available in varying sizes based in the number of teeth the pulley has. The pulleys come in various material types (i.e., aluminum, cast metal, and nylon) and have optional side guards called *flanges* to keep the timing belt centered on the pulley. Along with having a choice of pulley material types, you also have several options on the type of timing belt. These belts come single- and double-sided and even open-ended continuous (the two ends are clamped at points where they do not run over any pulleys). The belts also come in neoprene, urethane/polyester, and urethane/Kevlar, each having its own properties and tensile range.

Calculating the reduction ratio for this type of arrangement is easy and straightforward: you simply divide the number of the teeth on the output pulley by the number of teeth on the input pulley. For example, if you want a 3:1 reduction ratio, then one possibility would be to use a 60-tooth output pulley with a 20-tooth input pulley. Hence $60/20 = 3$. You will also need to know the centerline distance between the two pulleys to determine the proper-sized belt. That is where the use of an online calculator becomes invaluable. The one I generally use is located on the Stock Drive/Sterling Instrument Web site: http://sdp-si.com/. Realize that you will need at least one mounting plate (typically two) to mount the motor and shaft bearing—so calculating the centerline distance helps you to know how close you can put the two pulleys together. (Note that these are not like gears, the pulleys do not mesh!) It is also altogether possible to get the two pulleys so close to one another that they end up limiting the number of teeth engaging on the smaller pulley.

I have used all of the differing types of reducers over the years and find I favor the pulley reduction units the most. The reason is that these units are rather simple to build, are the least expensive, are very quiet during operation, and you can easily obtain the exact reduction ratio you are interested in. In addition (and this is a major contributing factor), there is no backlash to contend with as there is with a geared-type reducer, but it will take up more physical space. The downside to this type of unit is that you will most likely have to build it, as they are not commonly found for sale.

IMPORTANT NOTE ON PULLEYS: You will get a reduction if the drive pulley is smaller than the driven pulley. You will get an increaser if the drive pulley is larger than the driven pulley. So don't get them backward!

Constructing a Pulley-Reduction Unit

It is rather easy to make your own pulley-reduction units. Included here is a detailed example of how to construct one.

In this example, the goal is to achieve a 4.8:1 single-stage reduction driven with a NEMA 34-stepper motor, with a $\frac{1}{2}$-in output shaft for use with a pinion gear. Also desired is to use an XL or 0.200-in pitch belt with a $\frac{3}{8}$-in width. Using the calculator found on the http://sdp-si.com Web site, this ratio will work using 15- and 72-tooth pulley gears (i.e., 72/15 = 4.8).

Input inches as the units to work with in the calculator and select XL as the pitch. Give a desired ratio of 4.8:1. Select the "A" pulley material type as metal (plastic can be used) with 15 teeth. The "B" pulley is also metal with 72 teeth. Notice that there is a "center distance" value given that is reflective of the number of teeth of the belt that is chosen. The goal is to select a center distance long enough, or a belt with a high enough number of teeth such that both the A and B pulleys have an acceptable number of "teeth in mesh". Figure 3-7 shows a screen capture of the online calculator with these input values. A belt with 86 grooves and centerline distance of 3.8091 inches works well for this example. Also make sure the PD of pulley B (the large one) is just over 4.5 in. Hence, you can surmise that you will need a set of plates that are 5 in wide.

Using a CAD program, draw two circles side by side each corresponding to the PD of both the A and B pulleys, with a center distance between the two pulleys of 3.8091 in. Since the footprint of a NEMA 34 motor is 3.4 in square, draw a 3.4-in square centered over the small pulley. It is now easy to determine that a good length to make the mounting plates will be 8.25 in. For one reduction unit, you will need two $\frac{1}{4}$-in aluminum plates, each 5 by 8.25 in. This too is a good time to mark some locations for where the spacers or standoffs will be located that separate the two plates (i.e., the pulleys and belt will be located in between the two plates). The diameter of the holes you drill through the plates will be dependent upon the size of machine screw you use. You will also find it an advantage to drill the holes with the plates in pairs—that way everything will line up nicely when you go to assemble them. Note that the overlap showing between the large pulley and the motor

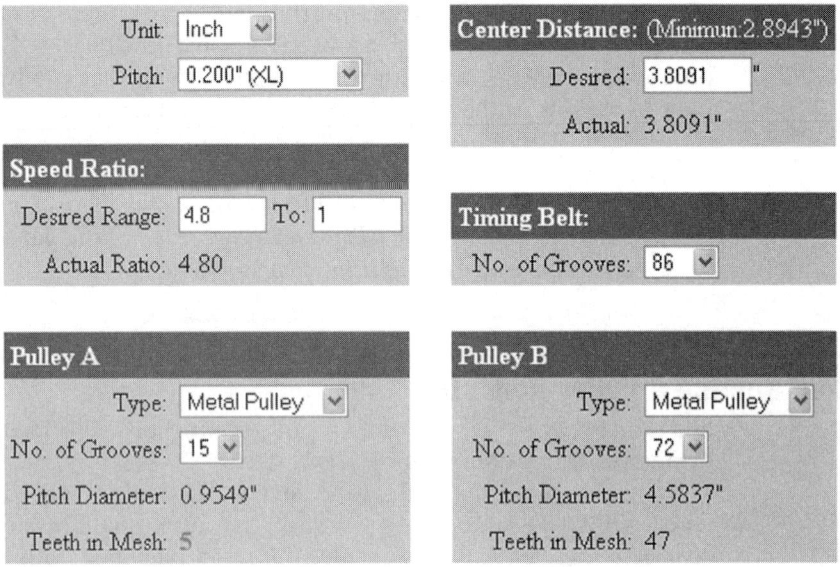

Unit:	Inch	Center Distance:	(Minimum: 2.8943")
Pitch:	0.200" (XL)	Desired:	3.8091"
		Actual:	3.8091"

Speed Ratio:
Desired Range: 4.8 To: 1
Actual Ratio: 4.80

Timing Belt:
No. of Grooves: 86

Pulley A
Type: Metal Pulley
No. of Grooves: 15
Pitch Diameter: 0.9549"
Teeth in Mesh: 5

Pulley B
Type: Metal Pulley
No. of Grooves: 72
Pitch Diameter: 4.5837"
Teeth in Mesh: 47

FIGURE 3-7   Online calculator.

footprints do not interfere with each other, as the motor will be on the outside of the plate and the pulley on the interior.

There will be a ½-in shaft driving the pinion. This same shaft will have the larger of the two pulleys clamped to the shaft—typically via set screws in the pulley's hub. This shaft needs to rotate as freely as possible, so we will use standard bearings to accomplish this. This type of radial bearing is commonly available either locally or by mail order. The Web site I recommend is http://www.vxb.com/ as I have been happy with both their price and quality. As each reduction unit has two plates, you will need two ½-in ID bearings and will most likely have an OD of 1.125 in. This is the diameter of hole size that needs to be made in both plates and centered with the large pulley. The plate that the motor gets attached to will also need a hole bored through it. This will allow the shaft of the motors to extend into the area in between the plates once they are assembled. I suggest boring a hole large enough for the small pulley to fit through for easier access. (In this example, I bored a 1.125-in-diameter hole, as the small pulley has an overall diameter of just under 1 in). Once the motor pulley hole is bored, you can install the small pulley on the motor and center it as best you can in the hole. Mark a location in the lower right-hand motor mounting hole for use as a pivot point. Bore and tap this hole and then reinstall the motor to the front plate. Now slightly rotate the motor both clockwise and counterclockwise, marking the three remaining motor-mounting holes in each position. After drilling and cleaning up the slots, you should now have a pair of plates that resemble those seen in Fig. 3-8.

FIGURE 3-8 Typical parts used in constructing a belt reduction unit.

Looking at the mounting face of your motor, you will notice a boss or large circular area that is raised from where the mounting surface is. If you have the facilities to produce a corresponding shallow area in the outside face of the front plate, go ahead and do so. However, it is just as easy to cut a spacer mounting ring in between the motor and the front plate that will take up the height distance of the motors' raised area. I generally make mine from 0.0625-in-thick aluminum sheet stock; the spacer does not need to have the slotted grooves. Finally, you use the motor itself to provide tension to the belt for proper operation by rotating the motor on the pivot point. In order to obtain the proper tension of the belt, it is recommended that you purchase and use a tensioning gauge, as that shown in Fig. 3-9.

Prior to assembly of the unit, you first need to countersink the mounting holes on the outside faces of both plates where the standoff spacers are located. This allows the back face of the unit to sit flush for mounting; use flush-mounting cap screws for your hardware. You should have all the pieces together for assembling your unit, as shown in Fig. 3-8. As a final step, you will want to place some type of material (typically thin sheet aluminum) around the perimeter of the two plates as a guard or safety shield. Not only will this keep cutting swarf from getting into the pulleys and belt during operation, but it will also keep fingers and loose clothing from getting caught up and causing a potentially hazardous condition. A set of completed reducers along with installed safety guards are shown in Fig. 3-10.

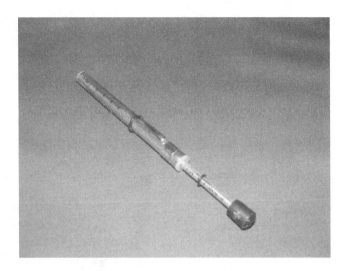

FIGURE 3-9 Belt tensioning tool.

Variations on the Pulley Reducer

There are a couple of other common variants that are used. The first is to make use of only one mounting plate. The surface mounting area of the bearing needs to be thick enough such that the shaft is sufficiently supported. This can be accomplished either by using one thick plate that is the thickness of the bearing or by using two plates with two bearings installed. One of the two plates will need to be the proper length and width, but the second need only be a smaller square or rectangular piece of material bolted to the larger plate in the area that houses the bearings. The overall thickness of the two plates should be such that they are as thick as the bearing being used.

FIGURE 3-10 Set of completed reducers and installed safety guards.

A second variant is a dual or multistage reduction unit. In this type of a configuration, you can achieve very high reductions. What this involves is placing a small pulley on the (would-be) output drive shaft. Via a second belt, this small pulley then drives a larger pulley that has the new output shaft for the pinion gear. Additional items needed in a multistage reducer will be an additional set of bearings, second shaft, two more pulleys, a belt, and tension idler for the secondary belt adjustment. This arrangement can be fabricated via two plates (as discussed above) or with the addition of a third plate placed in between the two outer plates to house the new set of output bearings. It is important to note that when adding a second (or third, etc.) stage output, the ratios are not summed, but rather multiplied together to obtain the final stage output. For example, if two 3:1 pulley combinations are used, the output is not 6:1, but rather 9:1.

CHAPTER 4
Motors

There are two types of motors that are typically used in CNC machinery: stepper and servo. Each of these motor types get both their power and direction/rotation information from a drive or amplifier. If you have spent any amount of time looking into the contrast of using a stepper as compared to a servo motor, you will find that there are distinct differences between the two. The basic operation of each type of these motors will be presented here for you to make your own determination regarding which type of motor would most benefit your system. After a discussion on each motor type, a contrast comparison of the two is presented. Hybrid configurations are also possible with the addition of an encoder to a stepper motor.

Stepper Motors

Undoubtedly the largest portion of hobby and midrange shop-based CNC machines available on the market today use stepper motors (Fig. 4-1). This style of motor has a rather large number of magnetic *poles* contained within the stator winding. Having a large amount of poles enables the rotor (the rotating part) to achieve very small increments of rotational movement. This ability is intrinsic to all stepper-based motors where each of the 200 divisions of rotation is 1.8 degrees. This is derived from 360 degrees of one rotation divided into 200 segments. These are the most common motors and, when coupled with a 10-step multiplier drive, the result is 2000 steps per revolution of the motor's output shaft. As a result of this intrinsic granularity, stepper motors do not *require* the use of an encoder for their operation and are considered *open loop* in this situation. Encoders are discussed in Chapter 5.

Stepper motors come available in many different physical sizes and adhere to an industry specification for standards known as National Electrical Manufacturers Association (NEMA). The specific sizes that are typically used on CNC devices, such as engraving machines, routers, and mills are NEMA 17, NEMA 23, NEMA 34, and NEMA 42. It is important to note that the smaller the number, the smaller the frame size is for mounting the motor. For example, the overall footprint for a NEMA 23 motor is 2.3-in square and 3.4-in square for a NEMA 34 motor. Stepper motors are available as single-, double-, or triple-stack, which relates to the amount of holding torque the motor is

58   The Physical Architecture

FIGURE 4-1 Various NEMA 23 and NEMA 34 motors.

capable of. Either the single- or double-stack configuration is used most typically, as they have lower inductance values and are capable of higher revolutions per minute.

The output or drive shaft for these motors will range from $1/4$ in on the smaller framed motors up to $1/2$ in or larger for the NEMA 42 size, although there is no standardization for diameters used.

Stepper motors can come with four, six, or eight pigtail wires, depending upon the way the windings in the motor were manufactured. The wiring configuration that makes use of six wires are referred to as unipolar. Both the four- and eight-wire leads are used in a bipolar wiring configuration. Within the eight-wire option, the user can choose to specify either bipolar parallel or bipolar series wiring. The difference in the parallel versus series option has to do mainly with the amount of current the motor will draw and the motors revolutions per minute capability. Wiring a stepper motor in bipolar series will yield the same rated holding torque, but will only require one-half the amperage to drive it and only achieve one-half of its potential speed. The motor wires are directly connected to the drives (typically located in the controller cabinet). The vast majority of drives available these days will be bipolar, meaning that there will be connections available on the drive for four wires. If you have a motor with either six or eight wires, you must follow the manufacturer's wiring schematic to achieve the four working wires. Unipolar or six-wire motors can be used as bipolar by simply isolating the two center tap wires – one from each of the coils.

Just by looking at a motor it is not obvious that there are connections or phases that are internal within the motor's wiring; hence, it is important which wires get paired off to each other when connecting them to the drive. However, if you have the wiring schematic and/or color codes for your motor this will simplify things. If you do not have this information, you can find the phases by touching the wires together to determine the phases. For example, if you had a motor with the following colored wires: red, blue, black, and white, by process of elimination and touching the red wire to the blue, black, and, finally, white, you will notice that the shaft of the motor will be more difficult to rotate by hand with one of these combinations. For the sake of this example, we will say that when the red wire touched the white wire, turning of the output shaft increased in difficulty. Testing by touching the blue wire to the black wire you should achieve the same result. You now know the wires associated with the two phases of the motor: red/white and black/blue. If you simultaneously short both of these phase wire combinations together, odds are you will not be able to rotate the shaft by hand. Connecting these wires to the drive would be as straightforward as the red/white wires going to "Phase 1" on the drive and black/blue to "Phase 2." Although the order is not critical, the pairing is. If you now apply a step signal to the drive, you will notice the motor shaft rotating. If it is not turning in the proper direction (meaning clockwise versus counterclockwise), you can simply swap the order of wires for just *one* of the phases. Going back to the same example, you could leave the order of the red/white wires, but reverse the second phase to be blue/black. This will result in a change in the direction of rotation of the motor.

A cautionary note must be made for nonstandard NEMA sizes.

It is well worth noting that there are stepper motors available on the market that do not adhere to the NEMA footprint sizing. The unit shown in Chapter 3 Fig. 3-5 is an example of a nonstandard NEMA footprint. Most times it is not a simple task to exchange out hardware parts, such as motors and motor mounts, when the mounting holes are different. Also, anytime you start having to deal with nonindustry standard equipment you can expect to have much less availability of sizing and options to work with, but you can also expect to pay much more money for something you may have to settle with in the first place. The recommendation is to always stay with hardware (and software) that is deemed industry standard. In the case of motors, staying with the NEMA frame series will save you some potential heartache in the future. It is much easier to exchange a NEMA 34 motor or reducer for another as compared to having to redrill or remake a motor mount bracket. Even the sale of your used equipment can be limited, as most buyers, in general, will be looking for standards-based equipment. If you are purchasing new equipment, ask the vendor these specific types of questions.

Typical specifications for a stepper motor could be shown in Table 4-1:

When people are new to stepper motors, most of the above information seems like gibberish to them. If you are purchasing a turn-key CNC machine,

Step Angle:	1.8 degrees
Coil type:	Two phase, unipolar six lead
Voltage:	3.2 V
Amperage:	2 A
Resistance:	1.8 Ω
Inductance:	2.5 mH
Holding torque:	250 oz-in

TABLE 4-1 Specifications for Stepper Motors

hopefully the company and/or person who designed the system will understand all of this information. However, choosing the right motor is not the only choice. In fact, what motor gets employed is directly connected to both to the stepper driver being used, the number of drives and motors that will be simultaneously used, as well as the capacity (voltage and amperage) of the power supply. In addition to these constraints, the user may have additional criteria, such as obtaining a certain cutting or feed rate, a desired rapid traversing rate, minimum cutting force, and resolution. One of the most important characteristics about a stepper motor is making use of its available rotary power (i.e., torque) within its speed range. Stepper motors all have their greatest amount of torque while spinning at lower revolutions per minute. There is a point in the speed of the shaft when the torque to rpm ratio begins to diminish. The point at which this happens is largely dependent upon the rating of the motor's inductance value. As a general rule of thumb, motors with a lower inductance value will tend to maintain their rated torque value at higher revolutions per minute before starting to decline. Understanding and making use of this concept is obviously more important in some applications as compared to others. Take, for instance, using stepper motors on a plasma cutting machine – there are no cutting forces to contend with. Hence, for both cutting and rapid velocity values, it is not as critical to maintain higher torque values than, say, those of a router or mill. In an application such as a router, you would like to ensure that your cutting force velocity (as equated to motor shaft rpm) is not beyond the point where the motor's available torque is dropping off. During rapid moves – the times when your spindle head is simply moving from point to point and not cutting – less torque is required and the only resistance forces to overcome are the frictional forces of the machine.

The motor's rated voltage is also an important value to understand with relation to working with a power supply. As a general rule of thumb, you want a voltage value that matches your DC power supply output that falls between 20 and 25 times the voltage value. For example, if the stepper motor is rated for 3.2 V, the range of power supply you should have driving it should

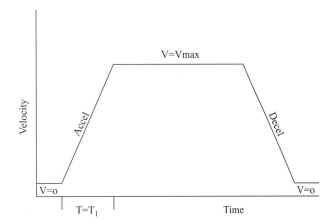

FIGURE 4-2 Velocity versus acceleration graph showing time T1 for acceleration to reach maximum velocity.

be in the range of 64 to 80 VDC. You must, however, ensure that the drive you are using also has the voltage capacity to handle this. In fact, to achieve both higher revolutions per minute and torque values you should select a power supply that comes as close to as possible (without exceeding) the stepper drive's rated voltage.

We conclude the discussion about stepper motors with one last bit of advice – bigger is not necessarily better and there is no "one size fits all" scenario. Be realistic and wise in choosing a motor that fits your application rather than using whatever is on hand or picking the largest one available. If you are unsure of which size to go with, solicit input from other users on the CNC Zone forum.

Determining the maximum performance and stability of a stepper system involves finding the point where the motors start missing steps. Once determined through testing, use of values in the controller software that limit both accelerations and velocities to 80 percent of the found breaking-point will typically ensure no loss of steps (i.e., position) during normal use. Figures 4-2 and 4-3 contrast the differences in acceleration values used in the motor setup and tuning for a generic machine. It must be realized that by using a lower acceleration value (as compared to the maximum), smoother starts and stops will be achieved. However, you do not want such a low value that it would affect the cutting speeds of the material(s) you are working with. Determination of both the maximum velocity and acceleration values are usually determined by the end user through a trial and error process.

Servo Motors

Servo motors and stepper motors can physically resemble each other when viewed from the outside. It's the configuration of the rotor and stator inside that makes the difference between the two. Servo motors in comparison to

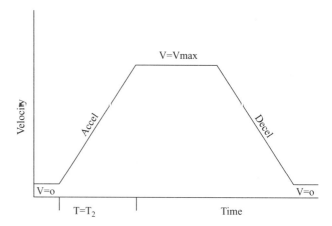

FIGURE 4-3 Velocity versus acceleration graph indicating a longer acceleration time (T2) to reach its maximum velocity.

stepper motors have considerably fewer number of magnetic poles in their windings. As a result, servo motors require the use of an encoder/resolver feedback device to be able to constantly make the required changes in order to achieve positional accuracy. This is what is referred to as "closed loop." In fact, without the use of positional feedback or closed-loop functionality, a servo motor would be useless for any type of CNC application. Most servo motors have a device known as a Hall-effect sensor that senses changes in the electromagnetic field of the motor during operation and is used to provide the feedback information to the drive/amplifier as to its exact physical location.

Servo motors have a much smaller rotor. The reasoning behind this is twofold: as the rotor is smaller and has less mass, much faster accelerations and decelerations are possible. The result is less time required to accelerate or ramp-up to the cutting speed or rapid traverse rates (i.e., the velocity). Second, higher rpm rates are possible. Unlike stepper motors, servo motors develop more power when operating at higher speeds.

When selecting a servo you will also have the choice of purchasing either a DC (direct current) or AC (alternating current) motor type. The DC type will be lower cost, but requires the use of brushes. Be aware that these brushes will wear out over time and will need to be replaced. AC servos, on the other hand, are brushless and require no maintenance and are the larger power producers of the two types. Also note that the motor type must match the drive/amplifier selection as well (i.e., AC versus DC). If you are interested in more of the specific details regarding the differences between these two types of motors or between an AC versus DC servo, an internet search will yield many responses.

Servo motors also require *tuning* for proper operation. There are two basic types of algorithms used to set their tuning parameters: PID (proportional integral derivative) and PIV (particle image velocimetry), with PID being the most commonly used. Additional tuning and tweaking for both frequency shaping and feed forward are also necessary. Discussions regarding which algorithm to use and the associated usage parameters go beyond the scope of this text. An Internet search on servo tuning will provide the reader a wealth of information on the topic. When purchasing new equipment that uses a servo system, you can rest assured that the manufacturer has already done the work involved in fine tuning the system.

Stepper versus Servo: Pros and Cons

This is an age-old debate: Which is better to go with, stepper or servo motors? Well, to be totally honest, there really is no clear-cut answer to this question as each has its own advantages and disadvantages; most of them are listed below. I will mention, however, that the bulk of hobby and lower- to mid-range production machines use a stepper-based solution. The reasoning behind this is typically a combination of lower cost and ease of implementation. In addition, if a CNC machine has been properly engineered to operate within the parameters that a stepper system has to offer, you should have no loss of steps or performance issues to contend with. Obviously, if something out of the ordinary were to occur, such as running your cutter bit into a clamp, there will be a probability that a loss of steps (loss of position) will occur. This is where the buyer or system engineer needs to make the determination to provide feedback or closed-loop functionality, or not. There is no doubt that a higher level of performance can be realized by using servo motors over steppers, but the cost difference can be appreciable. Ultimately, the decision for most users outweighs wants versus needs. If you are a first-time builder, I urge you to use stepper motors and drives the initial time around.

Advantages of Stepper Motor

- Lower cost – you will find that all of the components associated with stepper systems (i.e., motors, drives, etc.) are less costly.
- Very accurate and dependable under normal circumstances – intrinsic to stepper motors is their ability to achieve high positional accuracy.
- Optional 'hybrid' encoder feedback – if using dual-shaft motors, upgrading to use of encoders is a fairly simple task.
- No tuning required – other than operating just under maximum capabilities of your drive/motor, no tuning is necessary.
- Less mechanical reduction needed (generally) – as these motors operate best at lower speeds, lower reduction ratios are required.
- Simpler system to understand and work with – straightforward and easy to implement.

Advantages of Servo Motor

- Higher cost – motors, amplifier drives, and beefier power supply all come at a higher price tag.
- Closed-loop system provides the user 'insurance' – considered as a safety net.
- Higher mechanical reduction (generally) – the higher rpm rate may exceed the required application velocities needed and can be corrected via use of higher reduction ratio reducer.
- Tuning is required – trial and error testing required for proper system operation.
- Faster response and operation times – higher acceleration and deceleration rates.
- More complex to understand and troubleshoot – generally the case.

Encoders

When an encoder is used with a stepper motor it is commonly referred to as a *hybrid* configuration. Encoders are devices that optically, magnetically, or electronically monitor and provide feedback as to the *actual* position of each system axis in near real time. This information is then used to determine if the actual (i.e., physical) location, in fact, corresponds to the same position as what your G-code programming *intends* it to be. These units are typically used in conjunction with stepper motors that have dual shafts (see Fig. 4-4). Figures 4-5 and 4-6 show a complete encoder unit along with its internal components.

FIGURE 4-4
Stepper motor with dual-shaft configuration.

FIGURE 4-5
Encoder assembly.

Other common types of encoders are linear glass scales. These types are very precise and often used in applications such as mills and lathes. Note that it is cost prohibitive to purchase these types of scales in longer lengths, such as would be required on a router table with a larger working envelope. Glass scales are oftentimes optionally found on manually-driven metalworking equipment (mills, lathes, etc.) and the positional information is output to a digital readout (i.e., DRO) device. These scales are either positioned on the axis itself or on the machinery parallel to each axis.

FIGURE 4-6
Components of a rotary encoder.

PART II

The CNC Controller

CHAPTER 5
Controller Hardware

The physical controller is comprised of various components necessary in the conversion of software commands that result in electrical signaling to drive the motor. This chapter discusses the enclosure and various components that are used, as well as devices typically used in conjunction with the controller.

A rather important point I would like to stress is the *parallel port* connection (between the controller software on the PC and the hardware controller) is considered the industry standard. There may be some who promote use of either serial and USB connections and also claim that you will have a difficult time finding a computer with a db-25 parallel port on the PC to use. In reality, there are several companies that manufacture db-25 parallel port *cards* for use in your PC. In fact, you can purchase them in single-, dual-, quad-, and even up to eight ports per PCI slot. Hence, do not be alarmed if your computer does not already have a built-in db-25 parallel port. A standard card will cost between $10 and $30 (USD), and support up to 1.5 MB/s data transfers, which is more than adequate. All of the breakout boards listed in this chapter make use of a db-25 parallel port connection. The reasoning to avoid using a USB connection between the PC and the controller is because of latency. Although a USB connection has a higher data throughput, the time required for data connection establishment is significantly higher as compared to that of a db-25 parallel port. In addition, only small amounts of data are passed over the connection and parallel port bandwidth is more than sufficient. Therefore, it is advisable to avoid using controller breakout boards that employ a USB connection.

Enclosure

Figure 5-1 shows two examples of enclosures the author has designed and used in the past. These enclosures were formed up from 18-gauge steel and the through holes were made with a CNC laser cutter. Note that the use of a plasma cutter will also yield comparable results. In general, the goal of the enclosure is to provide a safe and dust-free environment for the electronics. It also provides easy access to wiring of the various components.

Properly sized access and mounting holes are usually made in the face of the housing for an ON/OFF switch, as well as an emergency power off (EPO)

FIGURE 5-1   Two examples of hardware enclosure cabinets.

switch. Typically the back of the box provides wiring access for the db-25 port and motor wiring connections. Other connections can include those for limit and homing switch input, Z axis zeroing touch-off, remotely controlled EPO switches, etc. Shown in Fig. 5-2 is an enclosure with a breakout board, five drives, spindle speed controller, and transformer ready for wiring.

Breakout Board

As the name implies, this type of printed circuit board performs a *breakout* of the signals present on each of the *pins* on the db-25 connection coming from the computer's parallel port. These boards are also commonly referred to as a BOB, which is short for breakout board. One of the primary functions of any breakout board is to provide some form of electrical isolation between the computer and the controller. This is critical as any anomalies in wiring or spikes in voltage will isolate the damaging electricity from migrating back to the physical port on the computer. These types of boards also provide convenient connections for the user to interface with the components. Some breakout boards have more features than others, but all provide basic connectivity required for step and direction signaling for each axis and phased output signaling for each motor – all via a db-25 input port. Note that the definition of *port* is generically used for the primary or secondary parallel port used on the controller's computer; pins are the individual wires on the port. In some instances, the use of an additional port is used to provide more input/output (I/O) functionality, as is the case in the use of a plasma torch height controller for the independent Z axis or if encoders are used in conjunction with stepper motors, etc. When more than one port is used, these are denoted by ports 1 and 2.

Controller Hardware **73**

FIGURE 5-2 Enclosure with a layout of components, ready for wiring.

Breakout boards come in many sizes, shapes, and abilities and most are typically designed for four motor/axis control. In slaving situations, where two motors drive one gantry, two motor outputs on the breakout board are required.

The board shown in Fig. 5-3 incorporates everything required for a power supply (minus the transformer) as well as a 120-V 15-A output relay to control a router head, dust collector, etc. It also provides output to drive up to five motors. A fifth motor output is nice to have in cases where a rotary axis is used on a router table. A notable feature of this board is that it directly incorporates motor wiring connectors. This approach requires no additional motor output wiring, as access to the connectors is available through the back of the cabinet. Note the board's db-25 connector is also accessible through the back of the controller.

Most boards are designed for four-drive axis control (see Fig. 5-4), whereas others have the ability to breakout signals from two parallel ports (Fig. 5-5).

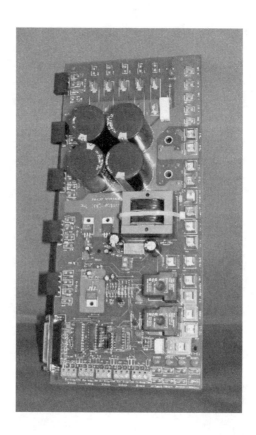

FIGURE 5-3 Five-axis multipurpose breakout board from Custom-CNC.

FIGURE 5-4 Four-axis BOB from PMDX with four Gecko 201 drives.

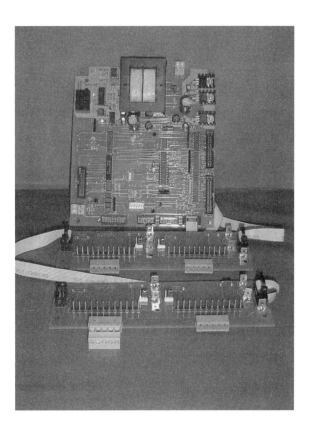

FIGURE 5-5 Dual-port BOB from Sound Logic.

Drives

Drives are devices that accept and amplify the signaling frequency rate coming from the breakout board. One drive is required per each motor to be used. Microstepping drives replicate the inbound step signals ten times for each one received. The drive multiplier takes a 200 step-per-revolution motor to yield 2000 steps/revolution, resulting in greater resolution.

In reference to Fig. 5-6, the units on the left and right are made by Gecko Drives, Inc. (http://www.geckodrive.com/) and are popular within the CNC hobbyist community. The drive shown in the middle of Fig. 5-6 is considered a high-performance stepper drive and is made by RTA Motion Control Systems (http://www.rtaeurope.com/rta/). During normal operation, drives produce a great deal of heat. As a result, heat sinks are used to provide cooling. Many drives provide intrinsic heat fins in their design (as in the RTA drive), while others require a user-supplied method for cooling. An aluminum plate is often used to dissipate the heat they generate.

Input connections to a drive are both step and direction signals along with output from the power supply. The circuitry within the drive amplifies these signals, which are then sent out to the coil windings in the motor. Note that

FIGURE 5-6 Some commonly available microstepping drives.

there are some drives that have individual power supplies incorporated into each drive where a standard A/C connection is made to each drive. In turn, each unit then rectifies the A/C input to provide D/C power for the drive's use. Obviously these types of drives are a bit larger than their counterparts, but then a power supply is not required in the controller enclosure.

Power Supply

All controllers will have some form of incorporated power supply (stand-alone or integral to the drives themselves) that provide the designed voltage and amperage necessary to drive the motors. Stand-alone regulated power supply units (or PSU's) come in various sizes and ratings (i.e., both voltage and amperage). Figures 5-7 and 5-8 provide some examples of stand-alone power supply units.

The user can also build their own power supply composed of individual components (see Fig. 5-9). Both rectifiers and capacitors are shown along with a transformer in Fig. 5-10. Optionally, a modular unit may be purchased that incorporates the rectifier(s) and needed capacitance in a convenient and easily mountable package. PMDX offers a power supply module (see Fig. 5-11) that only requires the addition of a transformer to complete the power supply unit.

There are many locations on the Internet that give the formulas used in component sizing, such as transformer, rectifiers, and capacitors, and are dependent upon the drive and motor to be used. The Web site http://campbelldesigns.net provides a good overview of the power supply component sizing and selection process. In any case, its capabilities must coincide with the overall design requirements of the system while maintaining both voltage and current limits of the drives. Note that the number of motors and

FIGURE 5-7 48-V 6-A power supply.

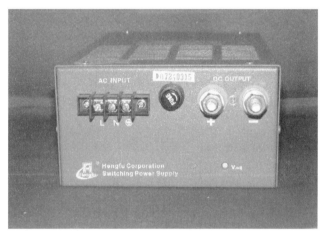

FIGURE 5-8 48-V 15-A power supply.

FIGURE 5-9 Power supply composed of individual components.

78 The CNC Controller

FIGURE 5-10 Torroidal-style transformer.

their individual voltage and amperage ratings will dictate the power supply requirements.

Adjunct Devices for Controller Hardware

There are many other adjunct devices used both in and with the controller. These are not necessary components used for basic motion control, but are, however, an overall part of the controlling hardware.

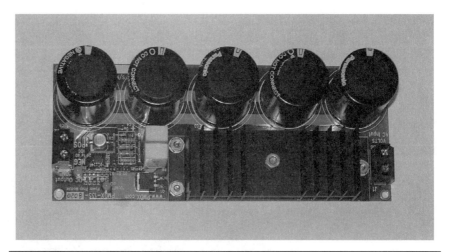

FIGURE 5-11 Power supply module.

Controller Hardware **79**

FIGURE 5-12 A dual-port BOB with spindle-speed controller from PMDX.

Spindle-Speed Controller

Most CNC applications that require rotary cutting or engraving have some type of a frequency-driven spindle. The use of a spindle-speed controller transfers the spindle control from the drives' panel to the software. The same functionality of the spindle (i.e., ON/OFF, forward/backward, and rpm) is then controlled via the computer with software commands. These commands can either be manually issued or from G code in an automated manner.

The circuit boards shown in Fig. 5-12 are controlled via the controlling software and output is to the variable-frequency drive. The card by Sound Logic (Fig. 5-13) is driven via a db-9 serial connection from the computer and does not take any pins away from a parallel port connection.

Pendant

A pendant is a relatively small hand-held device that the user plugs into either a USB or serial port connection, allowing remote control of the machine (obviously within the limits of the length of the cord). Keys on the pendant are able to be programmed or assigned various functions that are normally executed via the keyboard. Customary remote functions typically are: move, rapid, z touch off routine, pause, stop, file start, etc. There are several makes and models of pendants available on the market and most require the use of

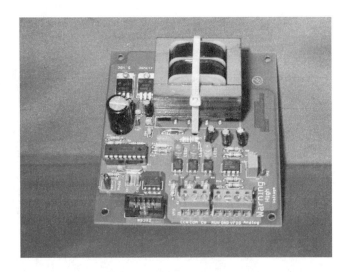

FIGURE 5-13
Serial-controlled spindle-speed-controller card from Sound Logic.

plug-in software to enable their use with the controller software. An example of one these devices is the Shuttle Pro from http://contourdesign.com; it is even possible to use your Xbox 360 controller (with Mach3). Here again, the user is responsible for determining which button on the controller gets assigned a specific function. The pendant the author prefers does not require a plug-in as it is a keyboard emulator and works with any controller software. This USB-controlled unit is from Texas Micro Circuits (http://www.texasmicrocircuits.com/) and a photograph of one is shown in Fig. 5-14.

In addition, you can build your own remote pendant. There are numerous free plans and examples available on the Internet you can use. Building your own enables you to custom design one for your specific needs. I would suggest

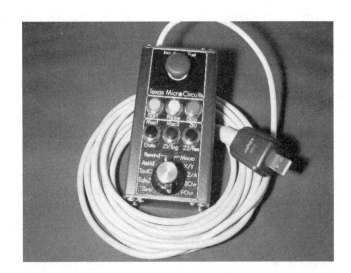

FIGURE 5-14
Remote pendant controller from Texas Micro.

Controller Hardware 81

FIGURE 5-15 A quality MPG from Nidec Nemicon.

that you use a quality manual pulse generator or MPG (such as shown in Fig. 5-15). This particular unit is from Nidec Nemicon Corporation, model #: OLM-01-2DZ1-11A.

Micro Switches

Micro switches can come in various sizes, shapes, and sensitivities and are used for physical determination of the maximum travel limits of each axis, as well as in homing (i.e., X/Y/Z zeroing) the application to a known starting point. Figure 5-16 shows the most commonly used type of microswitch and is mounted on a physically adjustable mounting plate.

FIGURE 5-16 A commonly used type of micro switch.

FIGURE 5-17 Set of magnetic switches.

These switches typically come with the ability of being wired as either N/O or N/C. For safety reasons, you will always want these wired as N/C.

Magnetic induction switches are another type of switch that could be used and function in a N/O configuration. In general, these types of switches should be avoided for the following reason: In a N/C scenario, a voltage signal is passed from the controller through all of the wiring and switches to ensure

FIGURE 5-18 Handy touch-off tool.

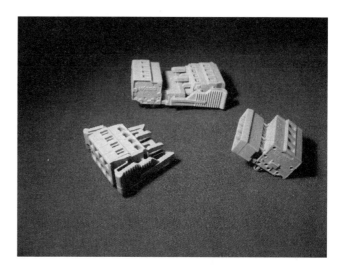

FIGURE 5-19
Commonly used WAGO connectors.

continuity is present throughout the circuit. If a N/O switch is used, there is no checking performed as the circuit is *open*. In the event a switch is faulty or a wire is broken, the discontinuity will not be sensed. This can result in a catastrophic failure, if the user is relying on the switch to provide a signal back to the controller. For safety reasons, always use switches that can be operated in a N/C mode. An example of a magnetic switch is given in Fig. 5-17.

Z Touch-Off Tool

Users can obviously construct their own tool to perform tool referencing from materials such as aluminum. The exact thickness of the tool needs to be known, as this dimension is subtracted from the measured distance. Users can also purchase an off-the-shelf tool that performs this function. The device shown in Fig. 5-18 utilizes a brass plate for the touch-off surface. Note that brass is a softer material than aluminum and can reduce the likelihood of damaging fine tool tips.

Wiring

Component wiring contained within the enclosure is typically nothing more than lengths of wire that have the insulation stripped off at the ends. In external wiring situations, where wiring needs to have some type of disconnect ability, snap or screw-together connectors are used. Figure 5-19 shows a widely used style of connector for stepper motor wiring. The example shown is four conductor and available in both male and female. Use of these is usually made at the controller box and at the individual motors.

CHAPTER 6
Control Software

There are several controller packages available to the user. Some are stand-alone software that support a wide range of hardware choices. Others are specific to work exclusively with a specific hardware controller, which would be considered a proprietary-based system and should be avoided if possible. Some are free whereas others almost require a second mortgage. As a user, your aim should be to find a controller package that is capable of interfacing and controlling hardware that is commonly available, as well as work with various peripheral devices. An Internet search with the key words "cnc controller software" will yield you a great number of choices. We will discuss the two most popular and powerful software controller choices currently available: Mach3 and EMC2. The type of input or machine-code language that both Mach3 and EMC2 use is also an industry standard (i.e., RD-274) and is referred to as G code. Later in this chapter, I provide various commands, both in tables and in detailed formats. Some CNC enthusiasts prefer to manually write their own G-code files; the use of an editor can greatly simplify the process. At the end of this chapter, we take a closer look at some G-code editors.

Mach3 Control Software

Most CNC users have either used, do use, or have at least heard of the Mach3 software (http://machsupport.com/). Without a doubt, it is the most popular controller software available. It costs $175 (USD) and is available with an optional CAM and milling add-on packages. The Mach3 software runs on a PC-controlled (via Microsoft Windows) operating system, can operate up to six axes simultaneously, and has considerable built-in functionality. This package supports user-defined Visual Basic macros and allows the user to enhance or completely rewrite the "front end", along with custom macros. This allows users to change and customize the look and feel of the application to suit their needs. Available on the Mach3 Web site are several dozen examples of modified screens that other users have created; all are available free of charge. There are also many user-written macro files that can be assigned to a new or existing screen button in the software, executed via command line or from G-code processing. This ability greatly allows each individual user to perform specific task routines that are custom to their operation.

Mach3 also allows the use of software *plug-ins*. When a hardware vendor wishes to add their product to the listing of Mach3-controlled devices, they can write plug-in software where the user simply downloads and adds it to their local copy of the software.

Via the use of a Wizard add-on package, the Mach3 software can also be used for conversational programming. This is where the various screens *converse* with the user for input to describe the part or cutting operation to be performed. Topics range from milling and thread forming on a lathe. Once the series of screens is complete, the software then outputs a file containing the necessary G-code descriptors.

Mach supports G code and comes prepackaged with many bells and whistles. You can obtain a free copy of the program that will support up to 1000 lines of G code directly from their Web site. This allows the user to take a test drive before purchase.

Enhanced Machine Controller, Version 2 (EMC2)

Although this controller software has been around for some time, it has not gained much popularity until rather recently. EMC2 is an Open-Source project that is written and supported by several software developers and is 100 percent free to anyone and everyone. The only "catch" to using this software is that it does not run in the Microsoft Windows environment. This program will only operate within the Linux operating system; see http://www.linuxcnc.org/ for more information. It will run in conjunction with any Debian-based UNIX distribution and is already available pre-packaged with an Ubuntu distribution of Linux: http://www.ubuntu.com/. If you are interested in this package, I highly suggest you download the version of Ubuntu that already has the EMC2 software incorporated from the linuxcnc.org Web site. (Note that this software is complied to only run under specific versions of Ubuntu.) As of this writing, the latest distribution is Karmic Koala (version 9.10). However, EMC2's latest release works with Hardy Heron (version 8.04 LTS). Users are able to download a "Live CD" version of Ubuntu/EMC2 and boot your computer from the CD ROM drive to preview the software and check their out its abilities.

Although EMC2 does not have all the bells and whistles of its counterpart, it does have some unique built-in features that others do not have. It uses a software shunt that is PLC programmable with ladder diagrams to aid in user-configurable tasking. One of its better features is the hardware abstraction layer (HAL) along with a built-in oscilloscope the user can use to graphically troubleshoot signal issues. It can help in the servo tuning process as well as stepper motor tuning and latency testing routines. EMC2 can also operate up to nine axes simultaneously and optionally uses kinematic modules to make use of other coordinate systems than the Cartesian system.

Users do not always tend to use EMC2 because of its cost. There are many applications that require higher feed rates (for example, step and direction signaling) than that EMC2 is able to provide. The reasoning behind this is because the computer is running the Linux operating system (a variant of UNIX) and has much faster processing ability as compared to Microsoft Windows. EMC2 is supported via an online forum, as well as an e-mail—based mailing list. Be advised that you will be required to issue some Linux commands via a terminal window from time to time. If you consider yourself to be computer illiterate, then the choice of EMC2 may not be for you.

A Foreword on Computer Operating Systems and Applications

These days, there are many choices of both operating systems (O/S) and applications to choose from. Undoubtedly the most common choice is Microsoft Windows with Macintosh and Linux coming in second and third, respectively. However, when establishing a work environment for CAD/CAM and controller software, it can be advantageous to work with the same O/S for networking purposes. This narrows down the choices to Windows or Linux. If you are a big fan of one or the other, stay with your favorite.

Windows is very common with a lot of help available in the event you run into trouble. It also has the most programs written that operate under it. The drawbacks with Windows are the cost of the operating system as well as the application software and controller software. Also, as Windows does not like to "let go" of its control over the CPU, a lowered number of step signals can be natively generated.

Linux, on the other hand, is open-source software and free of charge. Almost all application software that runs under Linux is free as well (for example, EMC2 controller software). A downside is that a fewer number of software packages are available (as in comparison to MS Windows) and many do not have as many features. A major advantage of Linux/EMC2 is its capability of very high rates in generating step signals. Hence, there is no need to purchase additional step-generator hardware, as is common with Mach3 users. For those who have never seen a computer running Linux, you may be surprised. Ubuntu (one version of Linux distribution) has a fully windows-driven graphical front end and looks very similar to Microsoft Windows.

If you are not pressed for time in getting your CNC machine into production or if you are looking for areas to reduce costs, I recommend you to look into running Linux operating system and preview some of the free software packages. If, however, you already have a computer running Windows, the low cost of purchasing the Mach3 software should not be a burden on your pocketbook.

G-Code Editors

When writing a G-code file from scratch or editing a file that has been prepared from a CAM program, an editor program can be very useful. They are available for use with both the Linux and Windows operating systems and are either free or available at very low cost.

One example of a G-code editor is AutoEditNC, which can be downloaded for free from http://betatechnical.com/autonc.htm. However, most editors work basically the same. This tool allows its user to enter commands directly, import a file, or make use of code insertion of canned G-code routines. It allows the selection of specified tooling to be used with the program, renumbering of the lines of code, and maintains a color-coded blocking and indentation for modal style of commands.

Where this type of software tool really excels is its ability to graphically simulate the cutting process without having to be on a computer with controller software. This allows the user to preview the process of the file prior to cutting and potentially not ruining a good piece of material by catching errors beforehand. For the beginner programmer, it's an especially good tool to ensure you are making proper use of either the G02 or G03 commands (clockwise or counterclockwise arcs). Yet another proper use of this tool is in writing and checking for errors of the main program calling one or several subroutines.

If you are the type of user that solely relies on the output of the CAM for cut files, you may not find this type of tool useful. If, however, you manually write your own G code or often modify the output of your CAM program then you will definitely find this type of tool beneficial.

G Code

The G code language is an alpha-numeric ASCII-based machine-command language that the controller interprets into discrete movements and modes. This is not the type of programming language that requires it to be compiled prior to use. However, it does need to adhere to a specific flavor or dialect that your controller can interpret and make use of. When configuring the output of a CAM package, you will be required to select a specific type of postprocessor file such that the resulting G code output will be understood completely by the control processor. Conversely, if you are writing your own file or issuing direct commands via the man data interface (MDI), all of the entries must be understood by the controller, else a syntax error will occur.

G code is considered the industry standard for machine tool-control language and its syntax is adhered to by a standard known as RD-274. There are a finite number of base commands that are intrinsic to the language and, depending upon the supplier of the controller software, there may be additional codes or parameters supported by their implementation. They will, however, all conform to the same basic command structure as is presented here.

It should be noted that there are a great many CNC users who never delve into manual programming or have even the most basic understanding of its syntax and use. It is altogether possible to solely rely upon the G-code output from your CAM post-processor to feed into your controller program. In this fashion, the user may lose some functionality that G code has to offer (for example, canned routines). However, it is a fast and easy method to produce the code output. There are also many other users who end up saving the output from the postprocessor and call the file manually via a handwritten G-code master file routine. Any of these methods, whether as individual or a combination, are deemed acceptable and are up to the individual user and their comfort level.

It is customary, and just good housekeeping, to preamble each of your G-code files with commands that instruct the file processing to be in the proper units (i.e., inch or millimeters), absolute or relative distances, etc. Such preamble strings are typically already established in the postprocessor file of your CAM software. Obviously for those manually coding their own files, these types of codes need to be specifically added to each file. Other good programming techniques, although not required, are the use of line numbers that precede each programming line or *block* of code. If there is an error in your code, the controller software will identify which line or lines it cannot process and indicate those specified line numbers.

It is not the intent of this book to provide an in-depth understanding of the G-code language and associated dialects. Beyond the basics covered in this text, the reader is urged to purchase the book "The CNC Programming Handbook, 3rd Edition" by Peter Smid for further information or if you are interested in manual programming techniques. The definitions, examples, and implementation of Peter's publication regarding RD-274 have become the standard if and when discrepancies occur. In addition, if you are interested in manual programming, a good bimonthly publication that incorporates real-world G-code examples is: http://www.digitalmachinist.net/.

There are numerous locations on the Internet where you can find both the G-code command listings along with their applicable definitions. I will caution you that many of these sources have either typographical errors or misinterpretations of the command usage and are not regularly maintained to reflect updates and/or corrections. The G-code listing contained here was obtained with the permission of Tormach, Inc. and is located online at: http://www.tormach.com/machine_codes.html. I suggest that you bookmark this page in your computer's browser for future reference to have an online copy available for use or to print out your own copy. Table 6-1 provides the most current listing and definitions of the available G codes available to date. The Tormach Web site provides examples of the applicable commands, showing their required syntax.

Summary of G Codes	
G00	Rapid positioning
G01	Linear interpolation
G02	Clockwise circular/helical interpolation
G03	Counterclockwise circular/helical interpolation
G04	Dwell
G10	Coordinate system origin setting
G12	Clockwise circular pocket
G13	Counterclockwise circular pocket
G15/G16	Polar coordinate moves in G00 and G01
G17	XY plane select
G18	XZ plane select
G19	YZ plane select
G20/G21	Inch/millimeter unit
G28	Return home
G28.1	Reference axes
G30	Return home
G31	Straight probe
G40	Cancel cutter radius compensation
G41/G42	Start cutter radius compensation left/right
G43	Apply tool length offset (plus)
G49	Cancel tool length offset
G50	Reset all scale factors to 1.0
G51	Set axis data input scale factors
G52	Temporary coordinate system offsets
G53	Move in absolute machine coordinate system
G54	Use fixture offset 1
G55	Use fixture offset 2
G56–58	Use fixture offset 3, 4, 5
G59	Use fixture offset 6/use general fixture number
G61/G64	Exact stop/Constant velocity mode
G68/G69	Coordinate system rotation
G73	Canned cycle – peck drilling
G80	Cancel motion mode (including canned cycles)

TABLE 6-1 G Codes

Summary of G Codes	
G81	Canned cycle – drilling
G82	Canned cycle – drilling with dwell
G83	Canned cycle – peck drilling
G85	Canned cycle – boring, no dwell, feed out
G86	Canned cycle – boring, spindle stop, rapid out
G88	Canned cycle – boring, spindle stop, manual out
G89	Canned cycle – boring, dwell, feed out
G90	Absolute distance mode
G91	Incremental distance mode
G92	Offset coordinates and set parameters
G92.x	Cancel G92 etc.
G93	Inverse time feed mode
G94	Feed per minute mode
G95	Feed per rev mode
G98	Initial level return after canned cycles
G99	R-point level return after canned cycles

TABLE 6-1 *(Continued)*

G00 – Rapid Linear Motion

For rapid linear motion (Figs. 6-1, 6-2, and 6-3), program: G0 X~ Y~ Z~ A~, where all the axis words are optional, except that at least one must be used. The G00 is optional if the current motion mode is G0. This will produce coordinated linear motion to the destination point at the current traverse rate

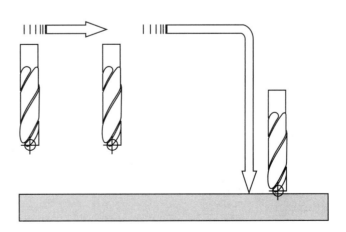

FIGURE 6-1 The G00 command is used to quickly move the tool from one point to another without cutting, thus allowing for quick tool positioning.

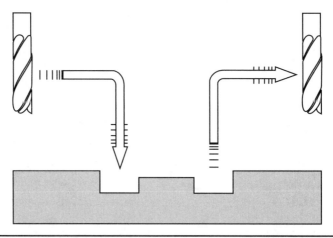

FIGURE 6-2 The rapid move of the G00 should have two distinct movements to ensure that vertical moves are always separate from horizontal ones. In a typical rapid move toward the part, the tool first rapids in a flat, horizontal XY plane. It then feeds downward toward the Z-axis. When rapiding out of a part, the G00 command always goes up the Z-axis first and then laterally in the XY plane.

(or slower if the machine will not go that fast). It is expected that cutting will not take place when a G00 command is executing.

If G16 has been executed to set a polar origin, then for rapid linear motion to a point described by a radius and angle G0 X~ Y~ can be used. X~ is the radius of the line from the G16 polar origin and Y~ is the angle in degrees measured with increasing values counterclockwise from the 3 o'clock direction (i.e., the conventional four quadrant conventions). Coordinates of the current point at the time of executing the G16 are the polar origin.

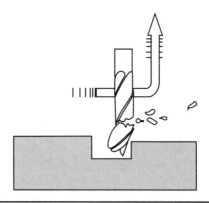

FIGURE 6-3 As this diagram shows, if basic rules are not followed, an accident can result. Improper use of G00 often occurs because clamps are not taken into consideration. Following the rules will reduce the chance of error.

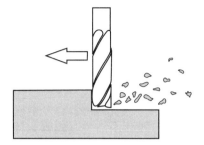

FIGURE 6-4 Linear motion, or straight-line feed moves on the flat XY plane (no Z values are specified).

If cutter radius compensation is active, the motion will differ from the above; see Cutter Compensation. If G53 is programmed on the same line, the motion will also differ.

Absolute coordinates depend where the tool is located. There are two basic rules to follow for safety's sake: If the Z value represents a cutting move in the negative direction, the X and Y axes should be executed first. If the Z value represents a move in the positive direction, the X and Y axes should be executed last.

G01 – Linear Motion at Feed Rate

For linear motion (Figs. 6-4 and 6-5) at feed rate (for cutting or not), program: G01 X~ Y~ Z~ A~, where all the axis words are optional, except that at least one must be used. The G01 is optional if the current motion mode is G01. This will produce coordinated linear motion to the destination point at the current feed rate (or slower if the machine will not go that fast).

If G16 has been executed to set a polar origin, then linear motion at feed rate to a point described by a radius and angle G00 X~ Y~ can be used. X~ is the radius of the line from the G16 polar origin and Y~ is the angle in degrees measured with increasing values counterclockwise from the 3 o'clock direction (i.e., the conventional four-quadrant conventions).

Coordinates of the current point at the time of executing the G16 are the polar origin. It is an error if all axis words are omitted. If cutter radius compensation is active, the motion will differ from the above. If G53 is programmed on the same line, the motion will also differ.

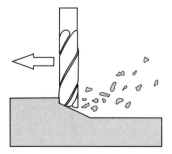

FIGURE 6-5 G01 command using multiaxis feed moves. All diagonal feed moves are the result of a G01 command where two axes are used at once.

96 The CNC Controller

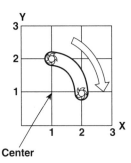

FIGURE 6-6 Shown is start point, endpoint, and center point of G02.

Note that because there is contact between the cutting tool and the work piece, it is imperative that the proper spindle speeds and feed rates be used. It is the programmer's responsibility to ensure acceptable cutter speeds and feeds.

G02 and G03 – Arc at Feed Rate

A circular or helical arc is specified using either G02 (clockwise arc) or G03 (counterclockwise arc) (Figs. 6-6 and 6-7). The axis of the circle or helix must be parallel to the X-, Y-, or Z-axis of the machine coordinate system. The axis (or, equivalently, the plane perpendicular to the axis) is selected with G17 (Z-axis, XY plane), G18 (Y-axis, XZ plane) or G19 (X-axis, YZ plane). If the arc is circular, it lies in a plane parallel to the selected plane.

If a line of code makes an arc and includes rotational axis motion, the rotational axes turn at a constant rate so that the rotational motion starts and finishes when the XYZ motion starts and finishes. Lines of this sort are hardly ever programmed.

If cutter radius compensation is active, the motion will differ from the above (see Cutter Compensation). Two formats are allowed for specifying an arc. We will call these the center format and the radius format. In both formats, the G02 or G03 is optional if it is the current motion mode.

In the radius format, the coordinates of the endpoint of the arc in the selected plane are specified along with the radius of the arc. Program: G02 X~ Y~ Z~ A~ R~ (or use G03 instead of G02). R is the radius. The axis words are all optional except that at least one of the two words for the axes in the

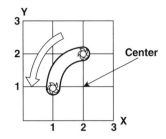

FIGURE 6-7 Shown is start point, endpoint, and center point of G03.

selected plane must be used. The R number is the radius. A positive radius indicates that the arc turns through 180 degrees or less, while a negative radius indicates a turn of 180 degrees to 359.999 degrees. If the arc is helical, the value of the endpoint of the arc on the coordinate axis parallel to the axis of the helix is also specified.

It is an error if:

- Both of the axis words for the axes of the selected plane are omitted;
- No R word is given;
- The endpoint of the arc is the same as the current point.

It is not good practice to program radius format arcs that are nearly full or semicircles (or nearly semicircles) because a small change in the location of the endpoint will produce a much larger change in the location of the center of the circle (and, the middle of the arc). The magnification effect is large enough that rounding error in a number can produce out-of-tolerance cuts. Nearly full circles are outrageously bad; semicircles (and nearly so) are only very bad. Other size arcs (in the range tiny to 165 degrees or 195–345 degrees) are OK.

Here is an example of a radius format command to mill an arc:

$$G17 \; G02 \; X \, 1.0 \; Y \, 1.5 \; R \, 2.0 \; Z \, 0.5$$

This means to make a clockwise (as viewed from the positive Z-axis) circular or helical arc whose axis is parallel to the Z-axis ending where $X = 1.0$, $Y = 1.5$, and $Z = 0.5$, with a radius of 2.0. If the starting value of Z is 0.5, this is an arc of a circle parallel to the XY plane; otherwise it is a helical arc.

In the center format, the coordinates of the endpoint of the arc in the selected plane are specified along with the offsets of the center of the arc from the current location. In this format, it is OK if the end point of the arc is the same as the current point.

It is an error if when the arc is projected on the selected plane, the distance from the current point to the center differs from the distance from the endpoint to the center by more than 0.0002 in (if inches are being used) or 0.002 mm (if millimeters are being used).

The center is specified using the I and J words. There are two ways of interpreting them. The usual way is that I and J are the center relative to the current point at the start of the arc. This is sometimes called incremental IJ mode. The second way is that I and J specify the center as actual coordinates in the current system. This is rather misleadingly called absolute IJ mode. The IJ mode is set using the button and LED on the settings screen. The choice of modes is to provide compatibility with commercial controllers. You will probably find incremental to be best. In absolute it will, of course, usually be necessary to use both I and J words unless, by chance, the arc's center is at the origin.

When the XY plane is selected, program: G2 X~ Y~ Z~ A~ I~ J~ (or use G3 instead of G2). The axis words are all optional except that at least one of X and Y must be used. I and J are the offsets from the current location or coordinates – depending on IJ mode (X and Y directions, respectively) of the center of the circle. I and J are optional except that at least one of the two must be used.

It is an error if:

- X and Y are both omitted;
- I and J are both omitted.

When the XZ plane is selected, program: G02 X~ Y~ Z~ A~ I~ K~ (or use G03 instead of G02). The axis words are all optional except that at least one of X and Z must be used. I and K are the offsets from the current location or coordinates – depending on IJ mode (X and Z directions, respectively) of the center of the circle. I and K are optional except that at least one of the two must be used.

It is an error if:

- X and Z are both omitted;
- I and K are both omitted.

When the YZ plane is selected, program: G02 X~ Y~ Z~ A~ J~ K~ (or use G03 instead of G02). The axis words are all optional except that at least one of Y and Z must be used. J and K are the offsets from the current location or coordinates – depending on IJ mode (Y and Z directions, respectively) of the center of the circle. J and K are optional except that at least one of the two must be used.

It is an error if:

- Y and Z are both omitted;
- J and K are both omitted.

Here is an example of a center format command to mill an arc in incremental IJ mode:

G17 G02 X 1.0 Y 1.6 I 0.3 J 0.4 Z 0.9

This means to make a clockwise (as viewed from the positive Z-axis) circular or helical arc whose axis is parallel to the Z-axis ending where $X = 1.0$, $Y = 1.6$, and $Z = 0.9$, with its center offset in the X direction by 0.3 units from the current X location and offset in the Y direction by 0.4 units from the current

FIGURE 6-8 The tool will pause for a short time (rarely more than several seconds). For a definite program pause, refer to commands M00 and M01. Being nonmodal the G04 command must be reentered each time dwell is to be executed.

Y location. If the current location has $X = 0.7$, $Y = 0.7$ at the outset, the center will be at $X = 1.0$, $Y = 1.1$. If the starting value of Z is 0.9, this is a circular arc; otherwise, it is a helical arc. The radius of this arc would be 0.5.

The above arc in absolute IJ mode would be:

$$G17\ G02\ X\ 1.0\ Y\ 1.6\ I\ 1.0\ J\ 1.1\ Z\ 0.9$$

In the center format, the radius of the arc is not specified, but it may be found easily as the distance from the center of the circle to either the current point or the endpoint of the arc.

G04 – Dwell

For a dwell (Fig. 6-8), program: G04 P~. This will keep the axes unmoving for the period of time in seconds specified by the P number. It is an error if the P number is negative.

G10 – Coordinate System Data Tool and Work Offset Tables

To set the offset values of a tool, program: G10 L1 P~ X~ Z~ A~, where the P number must evaluate to an integer in the range 0 to 255 – the tool number – and offsets of the tool specified by the P number are reset to the given. The A number will reset the tool tip radius. Only those values for which an axis word is included on the line will be reset. The tool diameter cannot be set in this way.

To set the coordinate values for the origin of a fixture coordinate system, program:

G10 L2 P~ X~ Y~ Z~ A~, where the P number must evaluate to an integer in the range 1 to 255 – the fixture number (values 1–6 corresponding to G54 – G59) – and all axis words are optional. The coordinates of the origin of the coordinate system specified by the P number are reset to the coordinate values given (in terms of the absolute coordinate system). Only those coordinates for which an axis word is included on the line will be reset.

It is an error if:

- The P number does not evaluate to an integer in the range 0 to 255.

If origin offsets (made by G92 or G92.3) were in effect before G10 is used, they will continue to be in effect afterward. The coordinate system, whose origin is set by a G10 command, may be active or inactive at the time the G10 is executed. The values set will not be persistent unless the tool or fixture tables are saved using the buttons on tables screen.

G12 and G13 – Clockwise/Counterclockwise Circular Pocket

These circular pocket commands are a sort of canned cycle that can be used to produce a circular hole larger than the tool in use or with a suitable tool (like a woodruff key cutter) to cut internal grooves for "O" rings, etc.

Program: G12 I~ for a clockwise move and G13 I~ for a counterclockwise move.

The tool is moved in the X direction by the value of the I word and a circle cut in the direction specified with the original X and Y coordinates as the center. The tool is returned to the center. Its effect is undefined if the current plane is not XY.

G15 and G16 – Exit and Enter Polar Mode

It is possible for G0 and G1 moves in the X/Y plane only to specify coordinates as a radius and angle relative to a temporary center point; program G16 to enter this mode. The current coordinates of the controlled point are the temporary center.

G17, G18, and G19 – Plane Selection

Program G17 (Fig. 6-9) to select the XY plane, G18 (Fig. 6-10) to select the XZ plane, or G19 (Fig. 6-11) to select the YZ plane.

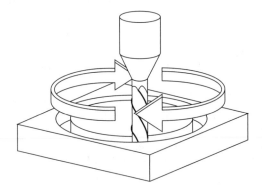

FIGURE 6-9 A circular tool move in the G17 plane.

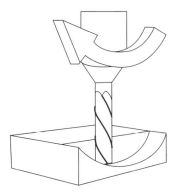

FIGURE 6-10 An example of an arc cut in the G18 XY plane. Bear in mind that because the primary and secondary axes are reversed, this arc is actually a G03 command.

G20 and G21 – Length Units

Program G20 to use inches for length units and program G21 to use millimeters. It is usually a good idea to program either G20 or G21 near the beginning of a program before any motion occurs and not to use either one anywhere else in the program. It is the responsibility of the user to be sure all numbers are appropriate for use with the current length units.

G28 and G30 – Return to Home

A home position is defined (by parameters 5161–5166). The parameter values are in terms of the absolute coordinate system, but are in unspecified length units. To return to home position by way *of* the programmed position, program: G28 X~ Y~ Z~ A~ (or use G30) (Fig. 6-12) . All axis words are optional. The path is made by a traverse move from the current position to the programmed position, followed by a traverse move to the home position. If no axis words are programmed, the intermediate point is the current point, only one move is made.

G28.1 – Reference Axes

Program: G28.1 X~ Y~ Z~ A~ to reference the given axes. The axes will move at the current feed rate toward the home switch(es), as defined by the

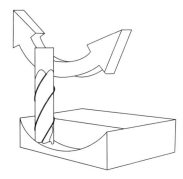

FIGURE 6-11 Tool cutting an arc in the YZ plane (G19).

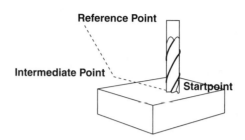

FIGURE 6-1.2 Cutter moves on the G28 command from the start point to the intermediate point and, finally, to the reference point.

configuration. When the absolute machine coordinate reaches the value given by an axis word, then the feed rate is set to that defined by configure>config referencing. Provided the current absolute position is approximately correct, then this will give a soft stop onto the reference switch(es).

G31 – Straight Probe

Program: G31 X~ Y~ Z~ A~ to perform a straight probe operation. The rotational axis words are allowed, but it is better to omit them. If rotational axis words are used, the numbers must be the same as the current position numbers so that the rotational axes do not move. The linear axis words are optional, except that at least one of them must be used. The tool in the spindle must be a probe.

It is an error if:

- The current point is less than 0.01 in (0.254 mm) from the programmed point;
- G31 is used in inverse time feed rate mode;
- Any rotational axis is commanded to move;
- No X-, Y- or Z-axis word is used.

In response to this command, the machine moves the controlled point (which should be at the end of the probe tip) in a straight line at the current feed rate toward the programmed point; if the probe trips, then the probe decelerates.

After successful probing, parameters 2000–2005 will be set to the coordinates of the location of the controlled point at the time the probe tripped (not where it stopped) or if it does not trip, to the coordinates at the end of the move and a triplet giving X, Y, and Z at the trip will be written to the triplet file if it has been opened by the M40 macro/OpenDigFile() function (q.v.). Code in macros or screen buttons can determine if a point is a trip or just the end of the move by inspecting if the DIGITIZE input is active after the G31 command or comparing the axis DROs with the requested move.

Using the straight-probe command, if the probe shank is kept nominally parallel to the Z-axis (i.e., any rotational axes are at zero) and the tool length offset for the probe is used, so that the controlled point is at the end of the tip of the probe:

- Without additional knowledge about the probe, the parallelism of a face of a part to the XY plane may, for example, be found;
- If the probe tip radius is known approximately, the parallelism of a face of a part to the YZ or XZ plane may, for example, be found;
- If the shank of the probe is known to be well-aligned with the Z-axis and the probe tip radius is approximately known, the center of a circular hole, may, for example, be found;
- If the shank of the probe is known to be well-aligned with the Z-axis and the probe tip radius is precisely known, more uses may be made of the straight-probe command, such as finding the diameter of a circular hole.

If the straightness of the probe shank cannot be adjusted to high accuracy, it is desirable to know the effective radii of the probe tip in at least the $+X$, $-X$, $+Y$ and $-Y$ directions. These quantities can be stored in parameters either by being included in the parameter file or by being set in a part program.

Using the probe with rotational axes not set to zero is also feasible. Doing so is more complex than when rotational axes are at zero; we do not deal with it here.

G40, G41, and G42 – Cutter Radius Compensation

To turn cutter radius compensation off, program: G40. It is OK to turn compensation off when it is already off.

Cutter radius compensation may be performed only if the XY plane is active.

To turn cutter radius compensation on left (i.e., the cutter stays to the left of the programmed path when the tool radius is positive), program: G41 D~. To turn cutter radius compensation on right (i.e., the cutter stays to the right of the programmed path when the tool radius is positive), program: G42 D~. The D word is optional; if there is no D word, the radius of the tool currently in the spindle will be used. If used, the D number should normally be the slot number of the tool in the spindle, although this is not required. It is OK for the D number to be zero; a radius value of zero will be used.

G41 and G42 can be qualified by a P word. This will override the value of the diameter of the tool (if any) given in the current tool table entry. It is an error if:

- The D number is not an integer, is negative, or is larger than the number of carousel slots;
- The XY plane is not active;
- Cutter radius compensation is commanded to turn on when it is already on.

The behavior of the machining system when cutter radius compensation is ON is described in the chapter on Cutter Compensation. Notice the importance of programming valid entry and exit moves.

Note: The tool offsets must have been applied with a G43 H~ for compensation to work.

G43, G44, and G49 – Tool Length Offsets

To use a tool length offset, program: G43 H~, where the H number is the desired index in the tool table. It is expected that all entries in this table will be positive. The H number should be, but does not have to be, the same as the slot number of the tool currently in the spindle. The H number may be zero; an offset value of zero will be used. Omitting H has the same effect as a zero value.

G44 is provided for compatibility and is used if entries in the table give negative offsets.

It is an error if the H number is not an integer, is negative, or is larger than the number of carousel slots.

To use no tool length offset, program: G49.

It is OK to program using the same offset already in use. It is also OK to program using no tool length offset if none is currently being used.

It is strongly advised to put the G43 command on the same line (block) as the T~ and the M06, which actually implements the change. If this is done then the control software anticipates the new offset during the time the operator has control for changing the tool. The operator can change the work Z offset safely if this condition is met.

G50 and G51 – Scale Factors

To define a scale factor that will be applied to an X, Y, Z, A, I, and J word before it is used, program: G51 X~ Y~ Z~ A~, where the X, Y, Z, etc., words are the scale factors for the given axes. These values are, of course, never themselves scaled.

It is not permitted to use unequal scale factors to produce elliptical arcs with G2 or G3.

To reset the scale factors of all axes to 1.0, program: G50.

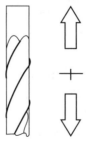

FIGURE 6-13 G50/51 – Scale factor.

G52 – Temporary Coordinate System Offset

To offset the current point by a given positive or negative distance (without motion), program: G52 X~ Y~ Z~ A~, where the axis words contain the offsets you want to provide. All axis words are optional, except that at least one must be used. If an axis word is not used for a given axis, the coordinate on that axis of the current point is not changed.

It is an error if all axis words are omitted.

G52 and G92 use common internal mechanisms in the CS and may not be used together.

When G52 is executed, the origin of the currently active coordinate system moves by the values given.

The effect of G52 is cancelled by programming: G52 X0 Y0, etc.

Here is an example. Suppose the current point is at $X = 4$ in the currently specified coordinate system, then G52 X7 sets the X-axis offset to 7 and so causes the X coordinate of the current point to be -3.

The axis offsets are always used when motion is specified in absolute distance mode using any of the fixture coordinate systems. Thus, all fixture coordinate systems are affected by G52.

G53 – Move in Absolute Coordinates

For linear motion to a point expressed in absolute coordinates, program: G1 G53 X~ Y~ Z~ A~ (or, similarly, with G0 instead of G1), where all the axis words are optional, except that at least one must be used. The G0 or G1 is optional if it is in the current motion mode. G53 is not modal and must be programmed on each line on which it is intended to be active. This will produce coordinated linear motion to the programmed point. If G1 is active, the speed of motion is the current feed rate (or slower if the machine will not go that fast). If G0 is active, the speed of motion is the current traverse rate (or slower if the machine will not go that fast).

It is an error if:

- G53 is used without G0 or G1 being active;
- G53 is used while cutter radius compensation is on.

G54 to G59 and G59 P~ – Select Work Offset Coordinate System

To select work offset #1, program: G54 and, similarly, for the first six offsets. The system-number–G-code pairs are: (1-G54), (2-G55), (3-G56), (4-G57), (5-G58), (6-G59).

To access any of the 254 work offsets (1–254), program: G59 P~, where the P word gives the required offset number. Thus, G59 P5 is identical, in effect, to G58.

It is an error if one of these G codes is used while the cutter radius compensation is on. (See relevant chapter for an overview of coordinate systems.)

G61 and G64 – Set Path Control Mode

Program: G61 to put the machining system into exact stop mode or G64 for constant velocity mode. It is OK to program for the mode that is already active. These modes are described in detail above.

G68 and G69 – Coordinate System Rotation

A rotation transformation can be applied to the controlled point coordinates commanded by a part program or by the MDI line. To do this, program: G68 X~ Y~ R~ The X and Y words specify the center about which the rotation is to be applied in the current coordinate system. R is the angle of rotation in degrees with positive values being counterclockwise.

If X or Y are omitted, then zero is assumed. A and B can be used as synonyms for X and Y, respectively.

To cancel rotation, program: G69, if a G68 is used while rotation is in operation, a G69 is implied before it. In other words, successive G68s are not cumulative and the X and Y points are always in an unrotated system.

When a rotation is in use, the X- and Y-axis DROs will be red to remind the operator that these values are program coordinate values that will be rotated.

This function can be used to compensate for work not exactly aligned on the table, to rotate the operation of a part program if it is coded with Y travel greater than X and so the work will not fit on the table or as software "vise soft jaws."

Note:

- G68 may only be used in the XY plane (G17 mode).
- The effects of changing work offsets when a rotation transformation is in effect will be nonintuitive, so it is wiser not to program this. Indeed, care should be taken when proving any program including transformations.
- There is very little standardization of the functions of this code across different CNC controls, so careful checks should be made on code written for other machines.
- Jogging always takes place in the direction of the machine axes. The toolpath display frame is oriented to the physical axes and will show the part at the angle at which it will be cut.

G73 – Canned Cycle – High-Speed Peck Drill

The G73 cycle is intended for deep drilling or milling with chip breaking. (See also G83.) The retracts in this cycle break the chip, but do not totally retract the drill from the hole. It is suitable for tools with long flutes, which will clear the broken chips from the hole. This cycle takes a Q number, which represents a "delta" increment along the Z-axis.

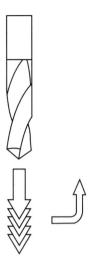

FIGURE 6-14 G73/83 – Peck drilling.

Program: G73 X~ Y~ Z~ A~ R~ L~ Q~

- Preliminary motion, as described in G81 to 89 canned cycles.
- Move the Z-axis only at the current feed rate downward by delta or to the Z position, whichever is less deep.
- Rapid back out by the distance defined in the G73 Pullback DRO on the settings screen.
- Rapid back down to the bottom of the current hole, but backed off a bit.
- Repeat steps 1, 2, and 3 until the Z position is reached at step 1.
- Retract the Z-axis at traverse rate to clear Z.

It is an error if the Q number is negative or zero. The following sample program demonstrates the G73 command.

G80 – Cancel Modal Motion

Program: G80 to ensure no axis motion will occur, to terminate canned cycles, etc. Note that it cancels the current G0, G1, G2, or G3 mode so this must be reestablished for the next move that is required. This particularly affects people adapting a CAM post-processor from another machine as this behavior varies between different CNC controls.

It is an error if:

- Axis words are programmed when G80 is active, unless a modal group 0 G code is programmed, which uses axis words.

G81 to G89 – Canned Cycles

The canned cycles G81 through G89 have been implemented, as described in this section. Two examples are given with the description of G81 below.

All canned cycles are performed with respect to the currently selected plane. Any of the three planes (XY, YZ, and ZX) may be selected. Throughout this section, most of the descriptions assume the XY plane has been selected. The behavior is always analogous if the YZ or XZ plane is selected.

Rotational axis words are allowed in canned cycles, but it is better to omit them. If rotational axis words are used, the numbers must be the same as the current position numbers so that the rotational axes do not move.

All canned cycles use X, Y, R, and Z numbers in the NC code. These numbers are used to determine X, Y, R, and Z positions. The R (usually meaning retract) position is along the axis perpendicular to the currently selected plane (Z-axis for XY plane, X-axis for YZ plane, Y-axis for XZ plane). Some canned cycles use additional arguments.

For canned cycles, we will call a number "sticky" if, when the same cycle is used on several lines of code in a row, the number must be used the first time, but is optional on the rest of the lines. Sticky numbers keep their value on the rest of the lines if they are not explicitly programmed to be different. The R number is always sticky.

In incremental distance mode: when the XY plane is selected, X, Y, and R numbers are treated as increments to the current position and Z as an increment from the Z-axis position before the move involving Z takes place; when the YZ or XZ plane is selected, treatment of the axis words is analogous. In absolute distance mode, the X, Y, R, and Z numbers are absolute positions in the current coordinate system.

The L number is optional and represents the number of repeats. $L = 0$ is not allowed. If the repeat feature is used, it is normally used in incremental distance mode, so that the same sequence of motions is repeated in several equally spaced places along a straight line. In absolute distance mode, $L > 1$ means "do the same cycle in the same place several times." Omitting the L word is equivalent to specifying $L = 1$. The L number is not sticky.

When $L > 1$ in incremental mode with the XY plane selected, the X and Y positions are determined by adding the given X and Y numbers either to the current X and Y positions (on the first go-around) or to the X and Y positions at the end of the previous go-around (on the repetitions). The R and Z positions do not change during the repeats.

The height of the retract move at the end of each repeat (called "clear Z" in the descriptions below) is determined by the setting of the retract mode: either to the original Z position (if that is above the R position and the retract mode is G98) or otherwise to the R position.

It is an error if:

- X, Y, and Z words are all missing during a canned cycle;
- A P number is required and a negative P number is used;

- An L number is used that does not evaluate to a positive integer;
- Rotational axis motion is used during a canned cycle;
- Inverse time feed rate is active during a canned cycle;
- Cutter radius compensation is active during a canned cycle.

When the XY plane is active, the Z number is sticky and it is an error if:

- The Z number is missing and the same canned cycle was not already active;
- The R number is less than the Z number.

When the XZ plane is active, the Y number is sticky and it is an error if:

- The Y number is missing and the same canned cycle was not already active;
- The R number is less than the Y number.

When the YZ plane is active, the X number is sticky and it is an error if:

- The X number is missing and the same canned cycle was not already active;
- The R number is less than the X number.

For preliminary and in-between motion, in the execution of any of the canned cycles, with the XY plane selected, if the current Z position is below the R position, the Z-axis is traversed to the R position. This happens only once, regardless of the value of L.

In addition, at the beginning of the first cycle and each repeat, the following one or two moves are made:

- A straight traverse parallel to the XY plane to the given XY position;
- A straight traverse of the Z-axis only to the R position, if it is not already at the R position.

If the XZ or YZ plane is active, the preliminary and in-between motions are analogous.

G81 Cycle

The G81 cycle is intended for drilling. Program: G81 X~ Y~ Z~ A~ R~ L~

- Preliminary motion, as described above.
- Move the Z-axis only at the current feed rate to the Z position.
- Retract the Z-axis at traverse rate to clear Z.

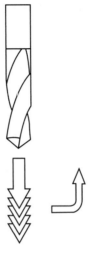

FIGURE 6-15 G81 – Canned Cycle Drilling.

G82 Cycle

The G82 cycle is intended for drilling. Program:
G82 X~ Y~ Z~ A~ R~ L~ P~

- Preliminary motion, as described above.
- Move the Z-axis only at the current feed rate to the Z position.
- Dwell for the P number of seconds.
- Retract the Z-axis at traverse rate to clear Z.

G83 Cycle

The G83 cycle (often called peck drilling) is intended for deep drilling or milling with chip breaking. (See also G73.) The retracts in this cycle clear the hole of chips and cut off any long stringers (which are common when drilling in aluminum). This cycle takes a Q number, which represents a "delta" increment along the Z-axis. Program: G83 X~ Y~ Z~ A~ R~ L~ Q~

- Preliminary motion, as described above.
- Move the Z-axis only at the current feed rate downward by delta or to the Z position, whichever is less deep.
- Rapid back out to the clear Z.
- Rapid back down to the current hole bottom, backed off a bit.
- Repeat steps 1, 2, and 3 until the Z position is reached at step 1.
- Retract the Z-axis at traverse rate to clear Z.

It is an error if:

- The Q number is negative or zero.

FIGURE 6-16 G83 – Canned Cycle Peck Drilling.

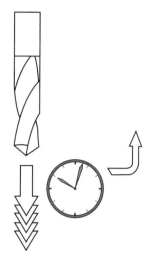

G85 Cycle

The G85 cycle is intended for boring or reaming, but could be used for drilling or milling. Program:
G85 X~ Y~ Z~ A~ R~ L~

- Preliminary motion, as described above.
- Move the Z-axis only at the current feed rate to the Z position.
- Retract the Z-axis at the current feed rate to clear Z.

FIGURE 6-17 G85 – Boring/Reaming Canned Cycle.

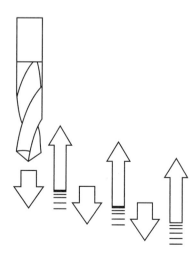

G86 Cycle

The G86 cycle is intended for boring. This cycle uses a P number for the number of seconds to dwell. Program: G86 X~ Y~ Z~ A~ R~ L~ P~

- Preliminary motion, as described above.
- Move the Z-axis only at the current feed rate to the Z position.
- Dwell for the P number of seconds.
- Stop the spindle turning.
- Retract the Z-axis at traverse rate to clear Z.
- Restart the spindle in the direction it was going.

The spindle must be turning before this cycle is used. It is an error if:

- The spindle is not turning before this cycle is executed.

G88 Cycle

The G88 cycle is intended for boring. This cycle uses a P word, where P specifies the number of seconds to dwell. Program: G88 X~ Y~ Z~ A~ R~ L~ P~

- Preliminary motion, as described above.
- Move the Z-axis only at the current feed rate to the Z position.
- Dwell for the P number of seconds.
- Stop the spindle turning.
- Stop the program so the operator can retract the spindle manually.
- Restart the spindle in the direction it was going.

G89 Cycle

The G89 cycle is intended for boring. This cycle uses a P number, where P specifies the number of seconds to dwell. Program: G89 X~ Y~ Z~ A~ R~ L~ P~

- Preliminary motion, as described above.
- Move the Z-axis only at the current feed rate to the Z position.
- Dwell for the P number of seconds.
- Retract the Z-axis at the current feed rate to clear Z.

G90 and G91 - Distance Mode

Interpretation of the CS-code can be in one of two distance modes: absolute or incremental. To go into absolute distance mode, program: G90. In absolute

Figure 6-18
Distance Modes –
G90 (absolute) and
G91 (incremental).

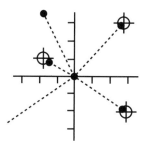

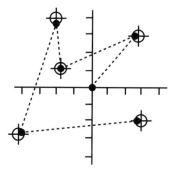

distance mode, axis numbers (X, Y, Z, A) usually represent positions in terms of the currently active coordinate system. Any exceptions to that rule are described explicitly in this section describing G codes.

To go into incremental distance mode, program: G91. In incremental distance mode, axis numbers (X, Y, Z, A) usually represent increments from the current values of the numbers.

I and J numbers always represent increments, regardless of the distance mode setting. K numbers represent increments.

G92, G92.1, G92.2, and G92.3 – G92 Offsets

You are strongly advised not to use this legacy feature on any axis where there is another offset applied.

To make the current point, have the coordinates you want (without motion), program:

G92 X~ Y~ Z~ A~, where the axis words contain the axis numbers you want. All axis words are optional, except that at least one must be used. If an axis word is not used for a given axis, the coordinate on that axis of the current point is not changed.

It is an error if all axis words are omitted.

G52 and G92 use common internal mechanisms in the CS and may not be used together.

When G92 is executed, the origin of the currently active coordinate system moves. To do this, origin offsets are calculated so that the coordinates of the current point with respect to the moved origin are as specified on the line containing the G92. In addition, parameters 5211–5214 are set to the X-, Y-, Z-, A-axis offsets. The offset for an axis is the amount the origin must be moved so that the coordinate of the controlled point on the axis has the specified value.

Here is an example. Suppose the current point is at $X = 4$ in the currently specified coordinate system and the current X-axis offset is zero, then G92 X7 sets the X-axis offset to -3, sets parameter 5211 to -3, and causes the X coordinate of the current point to be 7.

The axis offsets are always used when motion is specified in absolute distance mode using any of the fixture coordinate systems. Thus, all fixture coordinate systems are affected by G92.

Being in incremental distance mode has no effect on the action of G92.

Nonzero offsets may already be in effect when the G92 is called. They are, in effect, discarded before the new value is applied. Mathematically, the new value of each offset is $A + B$, where A is what the offset would be if the old offset were zero and B is the old offset. For example, after the previous example, the X value of the current point is 7. If G92 X9 is then programmed, the new X-axis offset is -5, which is calculated by $([7 - 9] + -3)$. Stated another way, the G92 X9 produces the same offset as whatever G92 offset was already in place.

To reset axis offsets to zero, program: G92.1 or G92.2. G92.1 sets parameters 5211 – 5214 to zero, whereas G92.2 leaves their current values alone.

To set the axis offset values to the values given in parameters 5211–5214, program: G92.3

You can set axis offsets in one program and use the same offsets in another program by programming G92 in the first program. This will set parameters 5211–5214. Do not use G92.1 in the remainder of the first program. The parameter values will be saved when the first program exits and restored when the second one starts up. Use G92.3 near the beginning of the second program. That will restore the offsets saved in the first program.

G93, G94, and G95 – Set Path Control Mode

Three feed-rate modes are recognized: inverse time, units per minute, and units per revolution of spindle. Program: G93 to start the inverse time mode (this is very infrequently employed). Program: G94 to start the units per minute mode. Program: G95 to start the units per revolution mode.

In inverse time feed-rate mode, an F word means the move should be completed in (one divided by the F number) minutes. For example, if the F number is 2.0, the move should be completed in one-half minute.

In units per minute feed-rate mode, an F word on the line is interpreted to mean the controlled point should move at a certain number of inches per minute, millimeters per minute, or degrees per minute, depending upon what length units are being used and which axis or axes are moving.

In units per revolution feed-rate mode, an F word on the line is interpreted to mean the controlled point should move at a certain number of inches per spindle revolution, millimeters per spindle revolution, or degrees per spindle revolution, depending upon what length units are being used and which axis or axes are moving.

When the inverse time feed-rate mode is active, an F word must appear on every line which has a G1, G2, or G3 motion and an F word on a line that does not have G1, G2, or G3 is ignored. Being in inverse time feed-rate mode does not affect G0 (rapid traverse) motions.

It is an error if inverse time feed-rate mode is active and a line with G1, G2, or G3 (explicitly or implicitly) does not have an F word.

G98 and G99 – Canned Cycle Return Level

When the spindle retracts during canned cycles, there is a choice of how far it retracts:

1. Retract perpendicular to the selected plane to the position indicated by the R word;
2. Retract perpendicular to the selected plane to the position that axis was in just before the canned cycle started (unless that position is lower than the position indicated by the R word, in which case use the R word position).

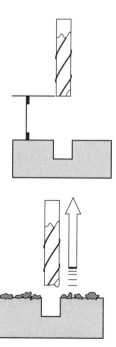

FIGURE 6-19
Z Retraction Level for Canned Cycle Operations.

To use option 1, program: G99. To use option 2, program: G98. Remember that the R word has different meanings in absolute distance mode and incremental distance mode.

M Codes

In addition to the numerous aforementioned listing of G codes, there are also a series of 'M' words that are used as well. These are termed 'M' as they are miscellaneous. A listing of M codes and their definitions are provided in Table 6-2:

M Code	Meaning
M0	Program stop
M1	Optional program stop
M2	Program end
M3/4	Rotate spindle clockwise/counterclockwise
M5	Stop spindle rotation
M6	Tool change (by two macros)
M7	Mist coolant on
M8	Flood coolant on
M9	All coolant off
M30	Program end and rewind
M47	Repeat program from first line
M48	Enable speed and feed override
M49	Disable speed and feed override
M98	Call subroutine
M99	Return from subroutine/repeat

TABLE 6-2 M Codes

M0, M1, M2, and M30 – Program Stopping and Ending

To stop a running program temporarily (regardless of the setting of the optional stop switch), program: M0.

To stop a running program temporarily (but only if the optional stop switch is on), program: M1.

It is OK to program M0 and M1 in MDI mode, but the effect will probably not be noticeable, because normal behavior in MDI mode is to stop after each line of input.

If a program is stopped by an M0 or M1, pressing the cycle start button will restart the program at the following line.

To end a program, program: M2 or M30. M2 leaves the next line to be executed as the M2 line. M30 "rewinds" the G-code file. These commands can have the following effects depending on the options chosen on the configure>logic dialog:

- Axis offsets are set to zero (like G92.2) and origin offsets are set to the default (like G54).
- Selected plane is set to XY (like G17).
- Distance mode is set to absolute (like G90).
- Feed rate mode is set to units per minute mode (like G94).
- Feed and speed overrides are set to ON (like M48).
- Cutter compensation is turned off (like G40).
- The spindle is stopped (like M5).
- The current motion mode is set to G1 (like G1).
- Coolant is turned off (like M9).

No more lines of code in the file will be executed after the M2 or M30 command is executed. Pressing cycle start will resume the program (M2) or start the program back at the beginning of the file (M30).

M3, M4, and M5 – Spindle Control

To start the spindle turning clockwise at the currently programmed speed, program: M3.

To start the spindle turning counterclockwise at the currently programmed speed, program: M4.

For a PWM or step/dir spindle the speed is programmed by the S word. For an ON/OFF spindle control it will be set by the gearing/pulleys on the machine.

To stop the spindle from turning, program: M5.

It is OK to use M3 or M4 if the spindle speed is set to zero; if this is done (or if the speed override switch is enabled and set to zero), the spindle will not start turning. If, later, the spindle speed is set above zero (or the override switch is turned up), the spindle will start turning. It is permitted to use M3 or M4 when the spindle is already turning or to use M5 when the spindle is already stopped, but see the discussion on safety interlocks in configuration for the implications of a sequence that would reverse an already running spindle.

M6 – Tool Change

Provided tool change requests are not to be ignored (as defined in configure>logic), The CS will call a macro (q.v.) M6Start when the command is encountered. It will then wait for cycle start to be pressed, execute the macro M6End, and continue running the part program. You can provide Visual Basic

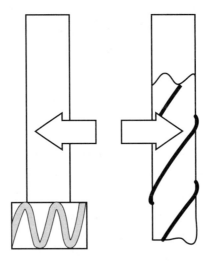

FIGURE 6-20 Tool Change using M6 Macro Command.

code in the macros to operate your own mechanical tool changer and to move the axes to a convenient location for tool changing, if you wish.

You are strongly advised to put the T~, the M06, and the G43 H~ on one line (block) of code. (See G43 for more details.)

M7, M8, and M9 – Coolant Control

To turn mist coolant on, program: M7.
 To turn flood coolant on, program: M8.
 To turn all coolant off, program: M9.
 It is always OK to use any of these commands, regardless of what coolant is ON or OFF.

M47 – Rerun from First Line

On encountering an M47, the part program will continue running from its first line.
 It is an error if M47 is executed in a subroutine.
 The run can be stopped by the pause or stop buttons.
 (See also the use of M99 outside a subroutine to achieve the same effect.)

M48 and M49 – Override Control

To enable the speed and feed override, program: M48. To disable both overrides, program: M49. It is OK to enable or disable the switches when they are already enabled or disabled.

M98 – Call Subroutine

To call a subroutine, program: M98 P~ L~ or M98~ P~ Q. The program must contain a letter O line with the number of the P word of the call (for instance,

O1, O125, O777). This O line is a sort of "label," which indicates the start of the subroutine. The O line may not have a line number (e.g., N123 O777) on it. The O line and the following code will normally be written with other subroutines and follow an M2, M30, or M99, so it is not reached directly by the flow of the program.

The L word (or optionally the Q word) gives the number of times that the subroutine is to be called before continuing with the line following the M98. If the L (Q) word is omitted, its value defaults to 1.

By using parameters values or incremental moves a repeated subroutine can make several roughing cuts around a complex path or cut several identical objects from one piece of material.

Subroutine calls may be nested. That is to say, a subroutine may contain a M98 call to another subroutine. As no conditional branching is permitted, it is not meaningful for subroutines to call themselves recursively.

M99 – Return from Subroutine

To return from a subroutine, program: M99. Execution will continue after the M98 which called the subroutine.

If M99 is written in the main program (i.e., not in a subroutine), then the program will start execution from the first line again. (See also M47 to achieve the same effect.)

Self-reversing Tapping Cycles

The P word specifies the depth to be threaded relative to the current Z position, which will typically be just clear of the work piece surface. The P word can be negative or positive with the same meaning.

Before use of these codes, the size of tapping head to be used and the pitch of the thread must be defined in the appropriated place on the settings screen. If the part program is running in inch (G20) mode, then the pitch is taken as a number of threads per inch. If it is metric (G21), then the pitch will be in millimeters. If the spindle speed is too high for the chosen pitch, then an error message will be displayed and the cycle will not be performed.

The cycle operates as follows: The currently set spindle speed and thread pitch are used to calculate the feed rate required to move the tap at the correct speed. The corresponding feed rate for the high-speed retraction of the tap is also calculated. If this exceeds the available rapid rate then an error is displayed.

The tap is then fed downward for the commanded depth (P word).

At the end of the down-feed, the spindle is rapidly retracted by the appropriate distance for the size of the head in use. This engages the reverse drive.

The spindle is then retracted, at the higher reverse rate previously calculated, for a distance sufficient to ensure the tap springs clear of the hole.

Letter	Meaning
A	A-axis of machine
B	B-axis of machine
C	C-axis of machine
D	Tool radius compensation number
F	Feed rate
G	General function (see Table 6-5)
H	Tool length offset index
I	X-axis offset for arcs X offset in G87 canned cycle
J	Y-axis offset for arcs Y offset in G87 canned cycle
K	Z-axis offset for arcs Z offset in G87 canned cycle
L	Number of repetitions in canned cycles/subroutines key used with G10
M	Miscellaneous function (see Table 6-7)
N	Line number
O	Subroutine label number
P	Dwell time in canned cycles Dwell time with G4 Key used with G10 Tapping depth in M871–M874
Q	Feed increment in GS3 canned cycle repetitions of subroutine call
R	Arc radius Canned cycle retract level
S	Spindle speed
T	Tool selection
U	Synonymous with A
V	Synonymous with B
W	Synonymous with C
X	X-axis of machine
Y	Y-axis of machine
Z	Z-axis of machine

TABLE 6-3 Other Codes

Order	Item
1	Comment (including message)
2	Set feed rate mode (G93, G94, G95)
3	Set feed rate (F)
4	Set spindle speed (S)
5	Select tool
6	Tool change (M6) and Execute M-code macros
7	Spindle On/Off (M3, M4, M5)
8	Coolant On/Off (M7, M8, M9)
9	Enable/disable overrides (M48, M49)
10	Dwell (G4)
11	Set active plane (G17, G18, G18)
12	Set length units (G20, G21)
13	Cutter radius compensation On/Off (G40, G41, G42)
14	Tool table offset On/Off (G43, G49)
15	Fixture table select (G54 – G58 & G59 P~)
16	Set path control mode (G61, G61.1, G64)
17	Set distance mode (G90, G91)
18	Set canned cycle return level mode (G98, G99)
19	Home, change coordinate system data (G10) or set offsets (G92, G94)
20	Perform motion (G0 to G3, G12, G13, G80 to G89 as modified by G53)
21	Stop or repeat (M0, M1, M2, M30, M47, M99)

TABLE 6-4 Order of Execution Table

The Z-axis is then positioned at the original height above the work, ready to move to another hole or another tool and operation.
Note:

- The above explanation is slightly simplified from the actual code used to aid understanding.
- For best results, especially for deep holes and blind tapping, the spindle speed chosen should be checked with a tachometer to ensure it is as near the commanded (S word) speed as possible.

M998 – Go to Tool Change Position

Execution of M998 will send the machine to the tool change position. The tool change position is defined on the settings screen. The Z-axis will move first,

The modal groups for G codes are
- group 1 = {G00, G01, G02, G03, G38.2, G80, G81, G82, G84, G85, G86, G87, G88, GS9} motion
- group 2 = {G17, G18, G19} plane selection
- group 3 = {G90, G91} distance mode
- group 5 = {G93, G94} feed rate mode
- group 6 = {G20, G21} units
- group 7 = {G40, G41, G42 } cutter radius compensation
- group 8 = {G43, G49} tool length offset
- group 10 = {G98, G99} return mode in canned cycles
- group 12 = {G54, G55, G56, G57, G58, G59, G59.xxx } coordinate system selection
- group 13 = {G61, G61.1, G64} path control mode

The modal groups for M codes are:
- group 4 = {M0, M1, M2, M30} stopping
- group 6 = {M6} tool change
- group 7 = {M3, M4, M5} spindle turning
- group 8 = {M7, M8, M9} coolant (special case: M7 and M8 may be active at the same time)
- group 9 = {M48, M49} enable/disable feed and speed override controls

In addition to the above modal groups, there is a group for non-modal G codes:
- group 0 = {G4, G10, G28, G30, G53, G92, G92.1, G92,2, G92.3}

TABLE 6-5 Modal Groups Table

then X and Y. An entry of 9999 will disable the axis. Execution of this function requires the machine to be referenced (homed).

User-defined M Codes

If any M code is used, which is not in the above list of built-in codes, then the control software will attempt to find a file named "Mxx.m1S" in the macros folder corresponding to the current XML profile name. If it finds the file, then it will execute the Visual Basic script program it finds within it.

New macros can be written using an external editor program like notepad and saved in the macros folder.

Modal Groups

Some of the G and M codes are modal meaning that the processor will remain in the same mode until superseded by another modal command (see Table 6-5). A classic example of this is the G00 or G01 command. If you (for example) issue a G00 command, all X-, Y-, Z-, and A-movement commands imply rapid or G00 movement. Once a G01 is executed, all subsequent X-, Y-, Z-, and A-axis commands will be of G01.

F – Feed Rate

To set the feed rate, program: F~. Depending on the setting of the feed mode toggle, the rate may be in units per minute or units per revolution of the spindle. The units are those defined by the G20/G21 mode. Depending on the setting in configure>logic, a revolution of the spindle may be defined as a pulse appearing on the index input or be derived from the speed requested by the S word or set spindle speed DRO. The feed rate may sometimes be overridden as described in M48 and M49 above.

S – Spindle Speed

To set the speed in revolutions per minute (rpm) of the spindle, program: S~. The spindle will turn at that speed when it has been programmed to start turning. It is OK to program an S word, whether the spindle is turning or not. If the speed override switch is enabled and not set at 100 percent, the speed will be different from what is programmed. It is OK to program S0; the spindle will not turn if that is done. It is an error if the S number is negative.

T – Select Tool

To select a tool, program: T~, where the T number is slot number for the tool. The tool is not changed automatically. It is OK, but not normally useful, if T words appear on two or more lines with no tool change. It is OK to program T0; no tool will be selected. This is useful if you want the spindle to be empty after a tool change. Note that if either a negative T number is used or a T number larger than 255 is used, an error will occur.

PART III
Application Software

CHAPTER 7
The Cartesian Coordinate System

Throughout the CNC process of CAD, CAM, and tool referencing, some type of coordinate system must be used to keep track of referencing. What takes place on the computer should match that of what occurs at the physical machine. Most readers will recall this coordinate system from one or more basic mathematics courses taken in the past. It is the Cartesian coordinate system that is typically used in the types of applications discussed throughout this book. There are other coordinate systems widely used in multi-axis robotics applications (kinematics, polar) that are not addressed in this text. If you are a person who has difficulty with math, please do not fret! You will not need to understand or compute any algebra or trigonometric equations. In fact, all that is involved is some very basic geometry.

Figure 7-1 shows a two-dimensional (i.e., 2D) representation of the Cartesian coordinate system. Particular points of interest are that the X and Y lines that are orthogonal (meaning, at 90 degrees to each other). Also note that where the lines of intersection cross is defined to have values of $X = 0$ and $Y = 0$.

Figure 7-2 shows the same 2D Cartesian coordinate system with a bit more information superimposed. Note that the intersection of the X and Y axes are the "zero" points for each axis, respectively. Hence, any values to the right of the Y axis will have a positive X value. Conversely, any values above the X axis will have a positive Y value. By definition, the Cartesian coordinate system is said to have four "Quadrants" (again, refer to Fig. 7-2). Arbitrary X and Y point values have been selected in each of the quadrants to show their corresponding X and Y numerical values. It is also customary to refer to the coordinate axes in a certain "order." That being X, then Y, and, finally, Z. There are actually several more possible axis designators, but we

FIGURE 7-1
Defining the coordinate system.

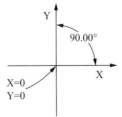

130 Application Software

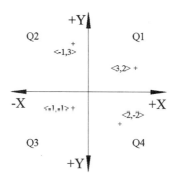

Figure 7-2
Coordinate system with quadrants and sample points.

will restrict our discussions to three dimensions. By using the X, Y, Z pattern it is not always necessary to denote the axis by its letter. A common shorthand method is to just use the numerical values for each axis while still adhering to the same pattern. For example, one way to represent the coordinate point of $X = 4$, $Y = 7$, and $Z = 2$, with the shorthand notation, would be <4,7,2>. Please refer back to Fig. 7-2 for various examples of this notation. Also pay particular attention to the fact that at least one of the X or Y values in quadrants II, III, and IV have at least one negative location value, whereas all coordinate values in quadrant I are positive values.

The most common physical topology of CNC-based implementations can be generalized by using either the mill or lathe architecture. From a generalized perspective to aid in the understanding of the Cartesian coordinate system, the term mill can be referred to as a metal working mill, but also a table router, plasma cutting table, water jet table, laser table, or even an XYZ pick and place machine. The thing that is common to all of the aforementioned systems is that they all are based upon three sets of orthogonally (i.e., right-angle) placed axes. The addition of the third axis is visualized as a line extending at right angles through the intersection of the XY plane and is at right angles to this plane. Positive Z values are determined from the right-hand rule. Using your right hand, extend your four fingers in the positive X direction. Then, curl your fingers into the positive Y direction. Your thumb will now be pointing in the positive Z direction of the coordinate system. There is more discussion on the Z axis later on.

The Table or Mill Topology

Although it is altogether possible to define the X and Y zero-points of your CNC tabletop as being directly in the center of the quadrants as shown in

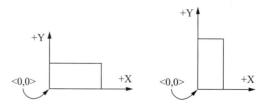

Figure 7-3
Showing typical XY zero locations on a table.

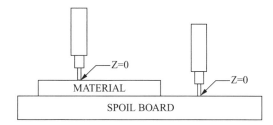

FIGURE 7-4 The two most common Z zero locations.

Figs. 7-1 and 7-2, it is universally accepted to adopt a "quadrant I" coordinate strategy. By defining the lower-left corner of the CNC work table as the $X = 0$ and $Y = 0$ being your reference point, this allows the CNC operator and programmer to work with positive values of both X and Y. (Please see Fig. 7-3.) (Important note: Although any point on the table can be defined as the zero reference point, we will adopt the "lower-left" corner as our X, Y zero reference point throughout this book unless mentioned otherwise.) Both rectangles shown in Fig. 7-3 are valid layouts as they both use the lower-left corner as a zero reference point.

Now that you are familiar with the two-dimensional coordinate system and the X, Y zero location as a reference point, all that remains in our simple 3D model is the addition of the Z axis. In practice, the zero reference point for the Z plane of motion is defined by the user (i.e., programmer and operator) at the time the machine code file is generated. Generally, there are two zero reference points used: The top of the material to be worked or the top of the table surface, usually referred to as the *spoil board*. It is important to note that the top of the spoil-board surface in actuality is the bottom of the material. Novice users typically have an easier time learning the concept of zeroing the Z axis (i.e., tip of tool cutter) at the top of the material to be machined. Hence, any Z axis values that are negative will be in the negative Z direction and will be cutting into the material. Conversely, any positive Z axis values will be above the material to be machined. (Please refer to Fig. 7-4 for a graphical representation of both Z zero points commonly used.) Just as knowing where your zero reference point is located in the X and Y planes, it is equally important to understand where your "Z zero" reference is located.

Lathe/Rotary Topology

By using the table or mill topology, we can now extrapolate this same coordinate convention system to describe a rotary or lathe topology. Referring to Fig. 7-5, you will notice that a basic lathe configuration consists of two

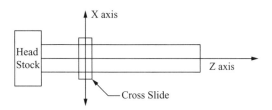

FIGURE 7-5 A top-down view of a lathe showing the axes.

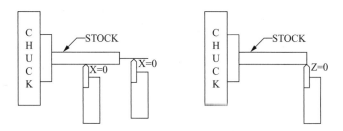

FIGURE 7-6 Lathe view showing possible X and Z zeroing locations.

orthogonal sets of axes (i.e., the X and Z). Unfortunately, there are not any typical and straightforward zeroing points used on the lathe as compared to the mill topology. The most generalized zero point along the Z axis is usually the point where the tip of the cutting tool abuts the end of the stock being turned. With regard to the X zero point, the two most common zeroing points are located either at the center line of the rotating stock or at the perimeter of the rotating stock. (See Fig. 7-6 for detail on the zeroing points located on a lathe.) Note that the lathe chuck is holding the rotating stock. It is customary to denote the longer axis that is inline with the spinning material to be the Z axis. The shorter axis is referred to as the X and it is placed at a 90 degree angle to the Z.

CHAPTER 8
CAD and Graphics

I would assume that almost all the readers of this text have already heard of the acronym CAD (computer-aided drafting). However, I will not assume that everyone has worked with it. CAD is the graphical process of drafting a part or an intended area, with vectors, in which CAM operators will be applied. It is most often used as the starting point in the "concept-to-part" process. CAD is extensively utilized in manufacturing industries throughout the world. Once the user is even somewhat proficient with a CAD package, a graphical representation of a part may be drawn (typically to a 1:1 scale) in a fraction of the time it would take to draw with pencil and paper. CAD also allows you to easily edit and make changes to the file without having to produce an entirely new drawing. CAD is quite simple to use and most software packages have built-in help facilities as a guide. In addition, there are often numerous texts and eBooks available that are specific to your brand of CAD, as well as courses available from the manufacturer or from your local college. As most CAD packages are capable of performing so many more functions than what you will be typically needing to draw your part, I would recommend you first "play" with your CAD package to familiarize yourself with the basics. Topics such as rendering and hatching, etc., are not required elements for CNC operations. If you are still uncomfortable, see if your CAD manufacturer has any additional free learning material that is available on their Web site or that they might mail to you. I usually find additional tutorial material to be very valuable as they contain many examples where they "walk" you through using most of the commands. Please bear in mind, that if 2D (i.e., two-dimensional) drawings will be your sole or even primary use of CAD, you will not need the latest and greatest nor most expensive version available. In fact, even the most basic of a CAD drawing package will provide sufficient features for the CNC user to accomplish any task.

What are some of the things that CAD can do? Consider the computer's screen background to be a canvas or grid that represents the first quadrant of a Cartesian coordinate system. This drawing grid typically is sized to either represent the piece of raw material you are milling your part from, or it represents the overall size of your machine's cutting area. Via pull-down menus or screen icons, the user has the ability to select any type of basic geometric shape and "draw" it anywhere on the grid. With CAD you do not

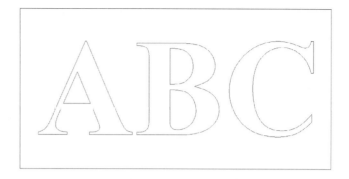

FIGURE 8-1
TrueType text.

have to scale down your drawings as is often done with the paper and pencil approach, which is typical to get all of the drawing all on one piece of paper. Hence, your drawing can be on a 1:1 scale and you can zoom in and/or out to view any portion of the drawing. You have drawing tool selections that will include a line segment, parallel lines, circles, arcs, ellipses, rectangles, line offsets, etc. You may also join segments together via "snap" commands so that there are no gaps or overlays in the part you are drafting – this is something that is very important as you will see in the CAM section of this chapter. Snaps allow the user to snap to the start and/or ending points of an existing arc or line segment, or an established grid point on the drawing canvas. All vectors have both starting and ending points and are referred to as nodes. It is critical in CNC machining that your line segments are all joined together, node to node.

Beyond the ability to draw your image from scratch, CAD programs allow the user to import other vector drawings, as well as apply text. The two types of text (font independent) will either be *engraving* or *stick font* text and *TrueType* text and are both are comprised of vector line segments. Examples of both types of text are shown in Figs. 8-1 and 8-2.

Pay particular attention to the difference between these two types of font types. In the TrueType example, each vector segment is bounded, meaning there are no open vector segments. Stick fonts, on the other hand, can have

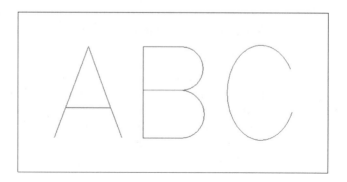

FIGURE 8-2 Stick font for engraving.

FIGURE 8-3
Example of pixelation.

both open or closed boundaries. You will notice this type of font used in engravings where the text is less than 1 in high and usually cut with conical engraving bits.

Before going any further, I would like to ensure that each reader understand why a *paint* type of program will not work in lieu of a CAD package. Drawing programs (such as Microsoft Paint) represent objects on the computer screen by illuminating *pixels* that are (loosely termed) "painted" onto the screen – one small square at a time. All images such as .jpg, .gif, and tif are all considered raster format and use pixelation to represent the object. If you use a paint program and zoom in, you will see jagged edges between the transitions between one pixel to the next. (An example can be seen in Fig. 8-3, where we zoom in on the center portion of a letter X.)

Vector files work on a completely different principle. Each and every vector consists of a discrete line segment and is represented by a starting and ending point, with a line connecting the two. These points can either be very close together or a large distance apart. Note that typical 2D vector files are easily also recognized by their file extension type: .dxf, .dwg, .ai, .eps being the most common.

Raster to Vector Conversion Utilities

Some CAD programs, as well as stand-alone software utilities, allow the user to import raster type images for conversion to a vectorized image. These are commonly referred to as raster to vector utility programs and there are many to choose from. This type of program traces the raster picture with vector lines and typically has threshold parameters to control the amount of smoothness the resulting trace will have. The author has used CorelTrace, which is by far the most powerful and easiest to use trace utility. It is available either as a stand-alone program or comes packaged with CorelDRAW. It allows object traces that are either inside, outside, or along the center line of the raster image.

If you are the type of user needing to convert photographs or any pixilated graphics, this type of tool will be very useful. A Web search will yield many such utilities that range from free to several hundreds of dollars. If you are interested in the freeware or shareware versions, they may be obtained from http://www.shareware.com/.

Difference between 2D and 3D

Both in this book and from other users, you will hear the term two-dimensional (2D) or 2½ D (another equivalent term) cutting versus three-dimensional (3D) cutting. There is a very basic distinction between the two.

- In 2D cutting, the cutter is moved to its intended cutting depth and held stationary while X, Y or both X and Y movement takes place. All CAM software is capable of performing 2D milling cuts. A classic example of this is where an item is profiled milled out of a section of flat stock.

- In 3D cutting, the cutter is allowed to move up and down while the X, Y or both the X and Y traverse through the material. This type of motion produces rounded contours in the material, without the need for round over or cove-style tooling. You will often see this type of milling used in sign work or dimensional facial artwork, etc.

Listing of CAD Vendors

Listed below are several of the most widely used CAD and graphics applications that are currently available. There are a wide range of not only cost, but types of features within those that are mentioned. It is of great importance for the novice to understand that an expensive program or one with more bells and whistles are not necessarily needed nor even desired! In fact, you will quickly realize that creating or editing a vector image only requires the most basic set of tools that any CAD graphics package has to offer.

AutoCAD and AutoCAD-LT

Without a doubt, AutoCAD is the world's leading PC-based CAD manufacturer. They have been around for a long time and you will find that other CAD programs will compare their software relative to AutoCAD as it is considered the *standard*. It is both "command line"-based as well as icon graphical, meaning that the user enters the various commands via the keyboard, or is able to just use the mouse. It seems to give the user the best of both worlds. I have used AutoCAD for many years in the past and have to admit that it is a very full featured program. It is capable of handling any 2D or 3D drawing task you may have. There are also more direct and aftermarket texts and tutorials available for this product than any others I have seen. They also have a 2D-only version of their product called AutoCAD-LT (light). Generally all you will need in most CNC work will be 2DCAD. The people at AutoDesk have also recently added a version that will work in a Macintosh operating system environment.

One drawback to using a command-line approach is that if you do not use it often enough to remember what the commands are, it takes the user a while to get back into the swing of things. Of course, you always have the "point and click"-icon driven method available – which is what most all other CAD manufacturers use by default. Another drawback for many users is the cost of these programs.

Currently, the initial cost for AutoCAD is $3,995.00 with upgrades from previous versions running from $595.00 to $1,795.00. The initial cost for AutoCAD-LT is $1,200. To view their package information and up-to-date pricing visit: http://www.autodesk.com

TurboCAD

TurboCAD is also a very popular choice among CAD users. It is a fully icon graphics-driven package that most find the easiest of all to use, and is considered to be the main contender to the suite of AutoCAD products. Most find that TurboCAD is easier for the novice to learn and operate, but it is often the difference in price that determines which of the two a user will purchase. In contrast, TurboCAD's high-end platinum version retails for $1,500. There are also versions available for those who use Apple Macintosh for their graphic workstation. The author uses this CAD package when needed and has found their user forum to be very helpful: http://forums.imsisoft.com. My particular use of this program, however, is not intended as an endorsement for readers of this book as I make use of this software in areas outside of CNC. Several years ago, the author purchased a used copy of TurboCAD from a bookstore for a considerably reduced price. I also purchased a very reasonable upgrade version. Most users of CAD for CNC purposes only utilize a small percentage of the package's overall ability, hence, even older versions of CAD are just as useful as an up-to-date version. The reader can see all of the IMSI products by visiting their Web page: http://www.turbocad.com

Free and Low-Cost CAD

There are many programmers around the world who have written their own CAD software package, and is available either as free or at very little cost. The main search engine to find this type of *freeware* or *shareware* is: http://www.shareware.com. There you will find literally dozens of CAD programs which can be downloaded. In addition you will find numerous other programs, such as CAM, raster-to-vector conversion tools, graphics utilities, etc., that can suit many of your needs.

QCAD

One program that can be found at http://www.ribbonsoft.com/ $31 USD for Windows or MacIntosh versions and free for Linux users is QCAD, which is a powerful 2D CAD package that is more than capable of handling pretty much any drawing task required of most beginning and intermediate CNC users. QCAD is my favorite program to use while working in a Linux environment.

Graphics Programs

Apart from the *paint* style of programs are several graphics software programs that have the ability to either work with or save the resulting output in a vectorized format. These programs are not marketed as CAD as their overall approach and use to drawing is more of a free-form style. Depending upon the type of application or, more specifically, your style of application will determine if a graphics style of program is better suited than a CAD package.

If your work is primarily within the signage industry, the use of a graphics program will allow you to produce a graphical layout (with color) to present to the customer. From this artwork, the desired vectors are easily extrapolated and saved for use with CAM software. There are many graphics programs available; however, not all of them have vector ability. For any not listed here, it is the reader's responsibility to determine prior to purchase if the program, indeed, has this ability and that the CAM program can import the output file prior to purchasing it.

Listing of Common Graphics Programs

- Adobe Illustrator – http://www.adobe.com/products/illustrator/ – very popular program for free-form artwork and extensively used in the signage industry. Very expensive.

- Rhino – http://www.rhino3d.com/ – used in several industries, such as, free-form artistry, metal machining, and mold production. Has optional CAM program. Very expensive.

- GIMP – http://www.gimp.org/ – excellent graphics program. Versions available that run in both MS Windows and Linux operating systems. The cost is free.

- SolidWorks – http://www.solidworks.com/ – has become an industry standard for use in rapid prototyping, mold making, etc. Very expensive.

CHAPTER 9
CAM Software

The decision to use CAM or not is largely dependent upon both the time and geometry of the item to be manufactured. Often, users either already have a CAD drawing or can rather quickly produce one to obtain any required dimensions in which to reproduce the item. The use of CAM affords the user to quickly yield a resulting tool path that, based on certain user-input criteria, can be used in the physical process. The required user inputs will include, but are not limited to, tooling used, depth per pass, plunge, and safe retraction of the Z-axis for moves in between cuts, feed rates for both cuts, and rapid moves, etc., all within the Cartesian coordinate system. Furthermore, all of the aforementioned parameters are output in an order of operations such that interior cutting functions are executed prior to peripheral operations, allowing the part being machined to remain affixed in place.

The user may also manually produce the G code – in essence by bypassing the use of a CAM tool. In this instance, it is the programmer's responsibility to individually denote the order of operations that would resemble those required to machine the same item if produced using a CAM software tool. The process sounds straightforward; however, for large files the task can be rather daunting. In addition, the G-code commands can either be saved into an ASCII text format for automated controller execution, or the same commands can be manually issued one at a time directly within the controller software via the man data interface (MDI). When this approach is taken, it is generally referred to as block processing.

There are yet other times when it can be advantageous to make use of both CAM and manual programming techniques. For instance, if the same item to be manufactured needs replication within a large sheet of material, the programmer can often save the CAM output in a separate file *naming* the file as a *subroutine,* thus denoting where on the sheet of material to repeat the same cutting operation. What this achieves is to greatly reduce the overall size of the cutting file by only providing the cutting parameters one time rather than multiple times in multiple locations. Please note that detailed examples demonstrating the use and operation of a main file calling a subroutine are covered in Chapter 11.

Understanding and Using CAM

Overall, the decision to use a CAM software tool or not depends upon the type of industry you are involved in. For instance, if you are producing 2D artwork for the signage industry, the use of CAM is almost a given. What would take an experienced G-code programmer hours (if not days) to produce an output file of a string of TrueType font letters nested on a sheet of material would only take minutes with a CAM program. Conversely if you are mainly doing 2D single-part file programming for mass production, it can be advantageous to manually write the necessary G-code commands to help streamline the cutting operation to yield the greatest number of widgets per hour of production.

It ends up being the users' responsibility to decide if they would benefit from using CAM versus a manual approach and also to decide when and where a combination of both would be worth the effort. To assist users in the manual only or combination approach, there are numerous G-code editing tool packages available to help the user in the manipulation of the code itself.

CAM, more so than CAD, is considered to be an intelligent piece of software. It accepts a vector-based graphical input file from which to work. The user then supplies the needed information, thus allowing the software to produce an output file with the properly ordered and formatted movement commands. CAM is an acronym standing for computer-aided manufacturing and is a software tool that establishes the cutting tool paths, feed rates, and outputs that result in a G-code format.

Each of the various methods available for use produce a G-code cut file that can have advantages associated with them – often based upon the size or quantity of the part(s), but, most importantly, the ease of obtaining the shape or *geometry* of the item you wish to manufacture. Bear in mind that absolutely anything *can* be manually programmed in G code, but not everything you may want to produce can be quickly and *easily* achieved by standard geometric shapes (i.e., circle, square, rectangle, hexagon) that manual programming can account for. To illustrate the comparison of the level of effort required between manual programming and automated CAM, consider a simple example, such as a simple square versus (for example) the letter "P" in a Times New Roman TrueType font (see Fig. 9-1). Visually, the reader can realize the differences in complexities between the two. From a high-level perspective, the square will have four straight sides to work from. The letter P, on the other hand, has both an interior and exterior vector path to traverse, along with transition curves or radii joining each of the horizontal and vertical connecting lines. Clearly, it is at these times that having a CAM program can make your life much easier when compared to manually attempting to input the necessary G-code commands.

For those who are not already familiar with what this software does for you, we will be going into some detail as to its purpose and examples of use. What the user inputs to CAM determines when and where the cutter bit is to

FIGURE 9-1
Profiling: simple versus complex.

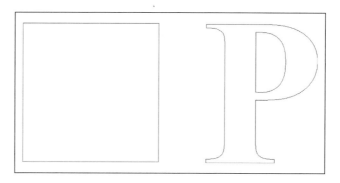

enter the material, traverse along the vector path, the direction of the path, and retraction for subsequent moves, along a safe path and height above the material. CAM automates these tasks (and more) with relatively minimal user input.

Overall, CAM provides the user the following:

- Reorder of vector line segments
- Check for *open* vector paths
- Establish tool paths
- Z-axis up/down for rapid traversing
- Start/stop and adjustment parameters for the application
- Provide G-code output file.

This summary of operations is discussed below in more detail.

The order of the line segments when drawn in CAD is something that is saved within the CAD file format. For example, if a box is desired and you initially only draw three of the four sides, then draw or edit somewhere else within the drawing and, finally, come back to draw the fourth side – all of these *order of operations* are recorded in CAD. Thus, CAM will *reorder* these vector line segments to make them contiguous or linked together. CAM also checks the drawing to see if there are any *open* vector paths and will typically denote any breaks with some type of a marker for user intervention. This allows the user to link together any unintended breaks in the artwork. Note that these types of errors can usually be avoided via the use of snaps while producing the artwork in CAD. Once the user has input the needed required data, CAM will produce what is referred to as a tool path. The location of the resulting paths are directly dependent upon the type of tooling the user has selected. For an application, such as routing or milling, the paths will be one-half the diameter of the chosen tool and typically located within or outside of the vector boundary. When producing a part that has both an inner and outer boundary or for multiple parts, the CAM program will establish plunge and retraction moves for the Z-axis, both for cutting and noncutting types of moves. This is required, as to not have the tool buried in the material when moving from location to location. Finally, almost all CAM programs accept user input that will be specific to the application. If using a high-frequency spindle head, for example, the user can select within CAM the tool number and the rpm to operate at. Thus, in the resulting output of the G-code file, there will be instructions to enable the spindle at the proper frequency and issue another command at the end of the file to turn off the spindle.

CAM Machining Parameters

In this section, we detail the common types of machining parameters found in CAM tooling setups. Please note that the terms describing the parameters that are listed here may or may not coincide with the same terminology used in your particular CAM package, but they will be close and perform giving the same end result.

Profiling – Inside and Outside

Profiling is the most basic type of milling or cutting operation. As the name implies, the resulting tooling path will profile (i.e., interior or exterior side) the vector boundary. Note that for this example we will restrict our discussion to use one router bit ($1/4$-in diameter or $1/8$-in radius) for all cutting operations. In the event that more than one tool is used, it is referred to as *rest machining*.

Figure 9-2 shows an example of a capital letter P in a Times New Roman font 4 in in overall height. The material to be used is $1/4$-in thick and is being cut from a large piece of sheet stock.

First, the reader should note that there will be two perimeter or profile cuts required to complete this operation. This is a very important aspect to the "order-of-operations" approach in CNC milling. As the part is being produced from sheet stock, the interior portion of the letter must be milled out prior to the outer portion. Consider if the outer profile of the letter were cut first, it would be much more difficult to retain the remaining material in its proper place on the spoil board to allow the interior profile operation to complete. Most, if not all, CAM programs have the built-in intelligence to

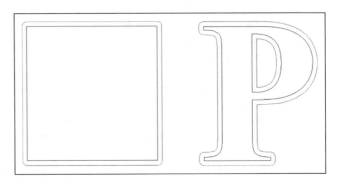

FIGURE 9-2
Sample profiling examples.

identify if a vector boundary does or does not have an interior boundary located or nested inside of the other. Once the $1/4$-in thick material (for example, Fig. 9-2) is placed on the spoil board and the cutter bit zeroed to the top of the material, the user would observe the following *generic* order of operations for this example:

- the bit will move to the start/stop point for the *interior* profile of the part
- the bit will move down into the material and stop at a $1/4$-in depth
- the bit will traverse along its cut path until it reaches the same start/stop point
- the bit will retract up to its determined "safe Z" height
- the bit will move to the *outer* perimeter start/stop point
- the bit will move down into the material and stop at a $1/4$-in depth
- the bit will traverse along its cut path until it reaches the same start/stop point
- the bit will retract up to its determined "safe Z" height
- the bit will return back to its $X = 0$ $Y = 0$ location

An important note is that this same overall order of operations will be generically the same for almost all CNC operations. Note these operations are very repeatable and predictable. Obviously the same operations would apply to the square shown in Fig. 9-2.

Area Clearance

This operation does just as its name implies. It clears out or removes material from a selected area that is bounded by a closed vector path. This function can be used to accomplish a host of differing results. Among these are pocketing a hole or pocketing an area that does not extend through the material. The area can be square, circular, or rectangular, flattening a work piece to a desired thickness, or removing only certain material while leaving other areas untouched, such as in milling a boss. A common example is when raised lettering is desired to be left on some material. The area clearance operation would be performed to all of the material where there is no lettering. Obviously you would not want to remove the full thickness of the sheet material as this would leave individual cut out letters. In essence, the lettering is left untouched and ends up raised, as compared to the surrounding areas.

The types of moves that are made to remove the material is usually user defined. There are two basic types of removal schema: raster or circular (Fig. 9-3). In raster, the cutter bit moves in one direction as far as possible and then advances a user-defined overlap distance and resumes cutting back along a parallel path to the previous one. Circular removal works basically the same, but with removal paths based upon the vector boundary shapes.

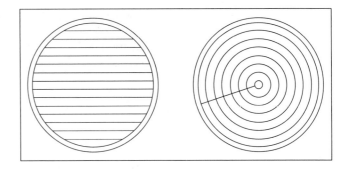

FIGURE 9-3 Area clearance using raster and circular options.

Cut Along a Line

This is a function that is often used in producing decorative types of cuts. Any type of router bit can be used – straight tipped or pointed. Often a ball-nose type of cutter bit is used to produce decorative fluting in millwork. Round-over bits used with this type of operation produce nice round-over profiles on stock as well. One of the most widely used times this operation is employed is during text engraving. The type of font that is used when the "cut along a line" option is enabled is referred to as a *stick* font (Fig. 9-4). Stick fonts differ from TrueType fonts in that they do not make use of closed vector boundaries. They are comprised of lines and arc segments that represent the shape of the letter.

Drilling Function

Your project will often require circular holes (Fig. 9-5) to be drilled or bored into your material in various locations. As with manual drilling, you too can have your CNC parameters set to do either through or blind drilling. The most important aspect to note is that the drilling function works only with

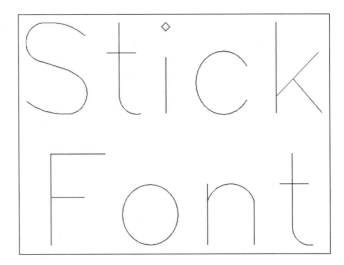

FIGURE 9-4 Engraving along a line using stick fonts.

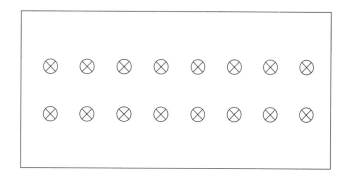

FIGURE 9-5 Hole drilling.

the diameter bit you are using. As an example, you have a piece of sheet stock that requires a series of ¼-in holes (possibly for shelf standards) along with some 35-mm holes for cup hinges. If you are using a ¼-in bit in your spindle, the drilling function will only plunge-drill the ¼-in holes to the proper size. To produce the 35-mm hole with the ¼-in bit, you would be required to select the area clearance option to produce the required circular pocket.

Inlay Function

Almost all CAM programs have the built-in ability for the user to apply an inlay of differing material to their work. What is required here is a "pocket" milled into your material, along with a cutout of differing material that matches the same dimensions and thickness of the pocket to fill the void. Even if your CAM software does not have this intrinsic feature, it is still easy to yield the same result. The steps are as follows:

- While the pocket is selected, perform an "offset" vector path command to the outside, by the diameter of the cutting bit you intend to use. Then select the newly offset path and issue the "offset" vector path command to the inside by the same diameter.
- While the inlay is selected, perform an offset to the inside by the bit's diameter.

Then perform another offset to the outside, again using same bit diameter.

Note that for this manual approach to work properly you will need to have the corner radius option in your software enabled throughout the offset procedures. This accounts for the radius or one-half of the diameter of the bit offset distance you entered.

The result you will now have is both a pocket and inlay piece that are of the same exact dimensions. Since they are exactly the same, the inlay will not yet fit into the pocket. The final step requires that you either offset the inlay piece (to the inside) by some *clearance* amount. Alternately, you can increase the size of the pocket with a path offset by the same small amount. The goal is to end up with a friction that fits in the inlay within the pocket. Trial and error test runs help the user determine what the proper clearance amount should be for the material type they are working with.

3D or Incised Carving

This type of carving is often used in conjunction with TrueType fonts that produce text engraved *into* your material at varying depths. During this type of engraving process you will have your X-, Y-, and Z-axis' simultaneously operating – hence, it is truly a 3D process. Note that the other milling procedures we have discussed maintain a constant Z-axis height throughout X and Y milling operations. This type of dimensional carving is obviously not limited to text, as it will work with any interior portion of a closed vector boundary (Fig. 9-6). It is, however, used exclusively with V-bit styles of cutters.

How this function works with a V-style cutter is completely dependent upon the varying width of the vector boundaries. For boundary areas that are closer together, the cutter tip will end up only slightly buried into the material. For wider areas, the cutter will extend deeper into the material – all the while maintaining a crisp V point at the bottom of the engraving. You can visualize that there is an imaginary center line shown that the cutter will travel along. This shows what the 2D cut path will look like from a bird's-eye view. What it does not show is what the height of the cutter will be during the

FIGURE 9-6
Rendered drawing of vector boundary showing full-depth V-bit engraving.

milling process that results in varying widths. Yet another method to envision the variable depth is to realize the angled sides of the V cutter will need to move into or out of the material in order to always span the width (i.e., vector boundaries) being carved.

This type of milling can also be performed by *limiting* the overall depth of cutting. In this case, the side walls or inside profile of the lettering is angled to match the angle of the V-bit cutter and the center of the area is cleared out flat rather than having an angled bottom.

With regard to the size of V-bit cutter, you should select one that has a cutter diameter or width equal to or larger than the widest point located within the boundary to be engraved. Another very important point in bit selection will be the *angle* of the cutter. For large letters, where the vector boundaries are fairly wide, it is common to use V bits of 120 or 150 degrees. The reasoning behind this is simple: you will often be limited by the material thickness. Using a steeper-angled bit would want to cut the V portion of the cut below the thickness of the material! There are, however, provisions to account for this by allowing you to limit the depth of cut to a desired amount. In the case where the cutter depth is limited, this process can take a rather extended period of time, while the V cutter "steps over" or moves a small incremental amount to remove the remaining amount of "flat space" within the boundary. To generalize, a steeper-angled V bit will require limiting the cut depth in contrast to the same vector cross section using a wider-angled V bit. Note that limiting the depth can achieve a pleasing aesthetic look without having to use a thicker piece of material.

If your CAM software has the ability to provide you with a 3D representation or isometric view of what the final engraving is to look like, you can *play* with different angled bits to achieve different results for the best look. Through time and experience, most users will knowingly be able to select the proper bit size and angle to fit the artwork's existing geometry. HerSaf is a brand of cutter I have used for many years, and I find them to be very accurate and dependable. They also make use of replaceable cutter inserts. New inserts are available for low cost and can even be resharpened by the user as needed. The angles of V-style cutters range from 60 to 150 degrees in 10-degree increments. (See their company URL for more information: http://www.hersaf.com/.) If you are into producing commercial signage, a complete set of these cutters will be invaluable in your shop.

Generalized Milling Options

Along with all of the types of milling operations (profiling, area clearing, etc.), CAM software will also require specific information other than just the diameter bit you intend to use. We will now cover most of the options that will be available via your CAM software, along with why and when you would want to use them.

Figure 9-7 Climb versus conventional milling.

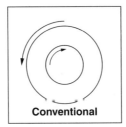
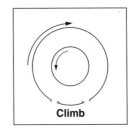

Climb versus Conventional Milling

This option tells your CAM output to have your cutter travel from left to right, or from right to left. This does not initially sound like such an important selection to make; however, it is of paramount in importance when CNC milling! If you have ever used a hand-held router before, odds are you are already familiar with what the proper direction of travel is: inside profiles go clockwise and outside profiles go counterclockwise. This orientation of direction is what is referred to in CNC terms as conventional milling. Then, what is climb milling? It's the reverse of conventional milling (Fig. 9-7). Again, if you are familiar with using a hand-held router, you may know this as back routing.

The reasoning behind having the ability to select either type of milling is simply a matter of how the finished edge of the material will appear. For example, most types of wood and wood composites will end up looking the best when a conventional direction is employed. Yet, there are many plastics and metals that look best when climb milling is used. Many of the bit-specific parameters you will need have already been derived for you. I have found the following URL to be very helpful: http://www.onsrud.com/ and select "Feeds and Speeds." From there you select the type of material you are working with and it will return a host of cut parameter information for your use. This vendor also has this URL: http://www.plasticrouting.com/ that provides additional information.

Note that the direction (clockwise or counterclockwise) will always be opposite of the other for interior versus exterior types of profiling.

An important notebook entry is the following: For any given material, you can cut a test square or circle. Make note of the direction in which you cut it (climb or conventional). For this example, let's assume your test cut in plastic was conventional. After cutting out the shape, examine both the edge of the piece that was cut out, as well as the edge of material that is left in the sheet of sample material. If the cut-out piece's edge looks best, then you know to use conventional milling. If, however, the cut in the sample section still on the cutting table looks better, odds are you need to be using climb milling. (Obviously if you are performing an operation, such as plasma cutting or straight-line font engraving, none of this would apply). Once you have figured out the milling strategy, jot this information down in your shop notebook. Later on you can add additional information such as spindle rpm's and feed rates to cut, based on cutting depth, etc.

Step-Over

This option is mainly used in conjunction with area clearing. Parameter options typically allow the user to indicate a specific incremental amount and also a percentage of step-over. Most users will initially want to have their step-over at its maximum for efficiency, but this is not always the case. Depending upon your machine, you stand the chance of leaving behind "fins" of unmilled material. Also, because of corner rounding, you too will leave behind material when performing square or rectangular pocketing routines. From experience, I tend to leave my step-over amount in the 50% range, but for another reason as well. Since your bit will only be removing one-half of its normal amount of material, you can greatly increase the cut feed rate as well as help extend your cutter's wear life.

Step-Down

Depending upon material type, cutter diameter being used, feed rate, etc., you will often find times when making a full-depth pass to cut your part is not achievable. A good example of this would be the cutting some $1/4$-in plywood versus cutting some 2-in-thick maple hardwood. This is where the cutter "step-down" option is used. Depending upon your CAM software, there will be differing options available for you to choose from. A general rule of thumb is to select a step-down amount that is equal to or less than the diameter of the cutter bit you are using. In the case of the 2-in-thick maple, if you have chosen a $1/4$-in bit for the task, it could take as many as eight passes for a profile cutout. In addition, there are times when the user wants the cleared area (during an area-clearance routine) that is left visible to end up with a finish that is as nice as possible. What can be done to achieve this is to perform a final clearance path that removes as little material as possible. This way the forces exerted on the cutter are minimized with little material to remove.

Another useful task of this feature is to only remove a thin amount of material on the final cut out pass. This helps reduce the cutting forces that are exerted on the part just prior to its removal from the sheet stock.

Finish Profiling

During repeated cutter step-downs on some material, slight ridges, which are undesirable, will be left visible on the profile. What the user can do to eliminate the stepped lines is to establish all the initial and/or step-down passes a very slight amount, oriented to the outside of the *final* vector path. Furthermore, the user should ensure that the final "roughing" pass does not completely cut through all of the material, but the part is left in place by a thin skin of material. The final pass of the cutter bit should then precisely cut up to the vector boundary line, as well as be set for a final depth to cut the part out.

Ramped Entry Cut

In order to cut out parts of a piece of material, the cutter bit needs to be "plunged" by some method into the material at the starting point (numerous times if using step-down). However, not all router bits and materials work well together during the straight-down plunging process. Setting up a *ramped* entry is many times the viable solution to help alleviate your ending up with an elongated hole that can show up in the final profile of the part. What this type of entry cut does is *saw* its way sideways parallel to the vector path until the desired step-down depth for the particular pass is met. This reciprocating action uses the cutter's side-milling ability to do the majority of the work, rather than just the tip of the cutter.

Lead In/Lead Out

This is yet another widely used option that can greatly help in eliminating the typical entrance/exit cutter marks that are produced from profile cuts. This feature allows the user to establish a bit entrance point that is some distance away from the actual vector path. Once the cutter is at the desired cutting depth, the bit is then lead into the associated vector path. Upon completing the milling of the part, the bit can then be optionally lead out of the vector path prior to being retracted. There also is typically an optional user-defined distance where the lead-out position will purposely overlap the initial lead-in position. This virtually eliminates profile marks left behind from plunging cuts. Note that this feature can also be used in conjunction with step-downs and cutter ramping.

Part Tabbing

Invariably there will be times when the user lacks the ability to safely hold the parts down to the table or it may also be desired to leave parts *slightly* held in place after the machining process is complete. Tabbing is a nice feature to use, which is especially important to stop your parts from flying off the table once the cutting is complete. How this is accomplished is by intentionally leaving behind small thin tabs of material after milling is completed. Once the milling operation is finished, the entire sheet of material can be removed from the spoil board and the parts "popped out." Quick sanding of the area where the tab was located is generally all that is needed to finish the part.

Nesting

Many CAM programs will allow the user to take a single (or multiple) part and replicate it multiple times on the size of the sheet material. In conjunction with this, the user can have the CAM program nest the parts to help minimize the overall amount of material needed. You will typically find options, such as number of times to replicate, along with spacing between the parts and the rotation origination. If you are working with a material that has a grain or working direction, the manner in which each part is oriented can become a key factor when being nested. If your work necessitates multiples of the same part to be cut from sheet stock, this is one feature you may find beneficial. It plays a major role in the amount of material ordered and in pricing.

CAD and CAM Combination Software

There are also several software packages available that the user can use for all of their drawing work (native or imported) and have the same package apply the tool paths as well. Obviously these packages will tend to be more expensive than their stand-alone counterparts. However, when considering both the overall cost of purchasing two independent software programs and the benefit of ease of use of working solely within one combination package, I tend to favor the all-in-one solutions. Obtaining support from one software manufacturer is also easier than having to deal with two or more and you won't run into the (rather common) issues of file version compatibility.

Listed below are some of the CAD and CAM packages I either use or have used in the past for various projects:

- Vectric Software – http://www.vectric.com/.
  This company has shown over the years to remain at the forefront of providing the users the greatest number of features while maintaining a relatively low cost. They also have a rather diverse offering of products. These range from 2D only and 3D engraving and routing, along with photomachining, and a huge library of vectorized artwork that the user can import for their use. What also helps set this company apart from the others is they regularly have in-depth training for user groups. During these educational meetings, they cover the program basics as well as the more advanced features, allowing the user to get the most from their package. In addition, Vectric has an online user forum to help with any questions that might arise.
- ArtCAM (by DelCAM) – http://artcam.com.
  If you are the type of user who primarily works with free-form types of designs, such as an artist or signage, you will find this package to be an advantage. Over the years this program has been the recipient of numerous awards for its innovativeness and relative ease of use. However, all of this comes with a rather hefty price tag, in some instances, costing the user as much or more than their CNC routers' price tag. They also have a great online forum and provide user training.

- MasterCAM – http://mastercam.com/.
  This combination package also works well with users who are free-form artistic, but also favors the type of user who is more machining concentric. They also pack a few more features into their software than most others with abilities for milling, routing, turning, and wire EDM, making this a very well-rounded purchase. If you are not, however, the type of individual or shop who requires all of the additional bells and whistles they have to offer, an equally powerful but use-dedicated piece of software might be for you.
- SheetCAM (by Stable Design) – http://sheetcam.com.
  This program initially started out by filling a low-cost powerful solution for users who do plasma cutting. Over the years, it has evolved into a universal package capable of complex designs and has grown to have specific user functions for milling, routing, oxyacetylene cutting, plasma cutting, water jet, laser, and more. Over the years I have submitted user enhancements that have been promptly incorporated. This package retails for $180 USD.
- CAM-Only Software
  Just as there are stand-alone CAD programs available, there are also CAM-only programs. This type of software allows the user to import any type of vectorized file. Once imported, you can establish the various types of normal machining operations that the CAM package has to offer. There are certain packages that handle 2D types of operations only, as well as some that include 3D cutting. Listed below are some examples of CAM-only software packages I have used in the past and were found to be both powerful and easy to use.
- VisualMill – by MechSoft: http://www.mecsoft.com/
- FreeMill – Free, but imports only VisualMILL, Rhino, STL, and VRML files. (Also by MechSoft Corporation.)

PART IV

Building or Buying a CNC Machine

CHAPTER 10
Choosing a Ready-Made CNC System

This chapter covers the selection of items that should be looked for when considering the purchase of a new CNC machine. In fact, if you currently have a machine that needs a retrofit or are considering the purchase of an older piece of machinery in preparation for a CNC conversion, most of these same selection items still apply.

Independent of the type of equipment you are interested in, here are a few key areas I highly recommend you adhere to. This applies to both new or retrofitting older equipment.

- Quality Construction

Don't always just go for the pretty looks! There are so many companies out there that have the same basic framework as the structural components of the machine, but have different appearances from one manufacturer to another. This is easily achieved by using thin sheet metal guards and panels that are often employed on the more expensive units. In many cases, you may end up spending from 50 to 100 percent more for the same basic unit and be paying for nothing more than some inexpensive addition of sheet metal.

In addition, you will find some companies who build systems from all extruded aluminum framing. This type of construction material is fine for the small-footprint systems (example 2 by 4 ft), but for larger table router systems there will not be enough mass and rigidity as compared to a welded-steel frame system. Other companies may try and use bent and stamped sheet metal for table frame and gantry components. This type of material does not work too well as it tends to flex and induce unwanted vibrations that end up resulting in the final cut in the material! The goal is having a solidly built and welded base with a lot of mass, along with a rigid, lightweight nonflexible gantry.

If you are upgrading an older machine (for example, a router table) and you are making use of larger motors, please realize that you will most likely need to either replace and/or stiffen existing framing members to handle the additional stress loads. If rebuilding, make sure you work toward having a stiff welded frame for a table. Anything bolted together is considered to be substandard and simply won't last. If the unit you are considering to upgrade had small torque motors installed and your upgrade is to put larger motors in their place, you can almost be assured stiffening of the old base or a new heavier frame will be in order. Larger motors are capable of producing much higher forces that can and will flex steel and aluminum members.

- G-Code Controlled System

The importance of using a system that operates in native G code is nearly as important as having a quality constructed machine! The reasons are numerous.

1. Any proprietary system (nonnative G code) will simply not have as many features as what the industry standard has to offer. Don't compromise.
2. Odds are you will purchase another CNC system in the future and it will more than likely be G code. Learn one coding language and stay with it.
3. If you pay someone to do your programming, G-code programmers are much easier to find than persons who may or may not know a proprietary language.
4. It is difficult to obtain books and programming examples for proprietary languages.
5. If your business expands, it's always best to have the machine operators using similarly controlled systems so operators don't have to be cross-trained.
6. You will receive less buyback money on a proprietary controlled system (unless you can somehow find someone who doesn't know what they are looking for!).

- Parallel port connection

For now and in the foreseeable future, your safest and best bet is to go with a parallel port connection between your PC and hardware controller. It is very tried and true technology and is not considered a bottleneck. Serial connections are simply too slow and the universal serial bus (USB) technology is not fully implemented and also suffers from high latency connection times. Also, for future upgrades and changes to other hardware vendor break-out boards, you will have the highest performing and lowest cost via the parallel port connection. Stay with the standard and save yourself a lot of heartache.

- Don't buy based on speed alone

Having the fastest CNC machine on the block is not necessarily always the best thing! For example, there are companies out there who will claim servo speeds using stepper motors. More often than not what they are doing is running the stepper's direct-drive rack and pinion without using any type of geared or pulley reduction. The result is high travel speeds, but low torque, resolution, and accuracy. When working with larger footprint machines, it is always advantageous to have higher travel rates, but do not compromise or the overall performance of the machine will suffer. If you truly need the higher speeds than what stepper motors have to offer, then invest into a servo-driven system.

- Controller software

As I have been mentioning throughout the book, it is not mandatory that you use a complete hardware and software solution through one vendor. If you are building your own hardware controller (based on parallel port) then you also have your choice of which controller software you like as well. There are many packages available, but I strongly urge you to use either Mach3 or EMC2. Mach3 does have several customizing advantages as you are able to build your own interface screens to suit your demands. Also, rather than running the native Mach3 interface screens, you can download for free many other screen sets from other users; Mach3 also supports a great many software plug-in modules for aftermarket peripherals. The software is very powerful, intuitive, and easy to use. The only word of caution I have to offer is it has more bells and whistles in features than you will probably need to exercise. So don't be dismayed the first time you see the software, thinking you need to know what each and every screen button is immediately used for. For beginners in CNC, Mach3 has an extensive worldwide support group who are eager to help with G-code programming, and even advanced milling techniques that span every aspect of CNC controls from routing and milling up to lathe turning.

- Motor power

Depending on your application, the amperage and torque rating of the motors can become a major factor in your decision making. If your main application is a large- format router table (4 by 4 ft or larger) and to cut sheet goods such as plywood, etc. then you want to choose a system that has larger steppers. Motors in the 4 to 6-Amp and 400–600 oz-in range are ideal for this type of cutting. In fact, I would not suggest going any smaller than a 3-Amp motor for any table in this size range. Of course, smaller router tables and engraving systems that do not have as wide a gantry system or cutting requirements can get by with smaller motors.

- Extras

There are obviously a great many other things to look for when choosing a CNC system, so pay close attention to the features that each has to offer. Most systems allow the user to turn ON and OFF the router head via the control software and G code (with a relay in the hardware controller), whereas others consider this an option or even want you to manually flip the switch on the router each time. If you plan to make use of a high-frequency spindle head, this too should be a low-cost upgrade option available to you. Via the use of a spindle head with a VFD, G code allows the user to not only turn the spindle ON and OFF, but to set the rpm as well. If the CNC system you are considering does not have provisions for these types of basics then you may wish to reconsider. Having to manually control your cutting head can become laborious. Also pay attention as to whether the computer that runs the controller software is included in the purchase price. If you are not a computer type of person that can perform the required tweaks to the, for example, Windows operating system, you should consider getting a computer already configured with all the software loaded ahead of time directly from the vendor. Another option is to purchase a system from a third-party vendor such as Warp 9 Computers (http://www.warp9computers.net). They provide new systems that are Mach3 ready with preloaded Windows XP. These computers already have the recommended tweaks enabling the user to get the highest performance out of the Mach3 application.

- User input

Getting input from a particular vendor's user group is advisable, but you should also be asking all of the pro and con questions of the system you are considering to an independently controlled and operated user group. I highly suggest you go to http://cnczone.com/ and either perform some searches or even post questions on your own. Here you will most likely get unbiased feedback.

Router/Plasma Table

There are several companies that are building and selling all size table routers. Most small and midsized companies are using the Mach3 controller software for their standard offering because of its wide acceptance and compatibility. If you are interested in purchasing, I recommend you opt for one of these companies. Other manufacturers are using paired package systems for their software controller and hardware. As I have been stressing all along – beware of proprietary systems and future upgrade compatibility issues.

Obviously I cannot cover all of the pros and cons of the numerous router and plasma vendors, but I will mention here the companies I either personally know to produce solid products and have great support. The vendors listed below obviously meet and exceed all of the selection criteria covered. This is not to be considered an exhaustive listing of vendors, but rather a good starting point for comparison purposes on things to look for.

- EZ-Router   http://ez-router.com/

EZ-Router has been around for several years and have captured a good portion of the market share for this range of router and plasma system. They have several systems in their standard product lineup, but excel at both frame and

Choosing a Ready-Made CNC System 165

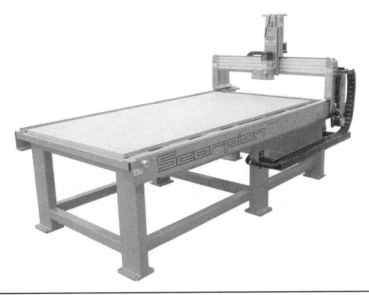

FIGURE 10-1   EZ Routers' Scorpion CNC router.

FIGURE 10-2   EZ Router - 4′ × 4′ CNC Plasma table.

gantry modifications to suit the customer's demands. Examples of customization are deeper gantry depths, longer and/or wider footprint tables, choice of linear guide systems, positional feedback for stepper-based systems, and servo upgrades. They also have chosen Mach3 for their controller software.

Listed below are three companies that offer excellent products (both routers and plasma machines) that I highly recommend.

- Tiger Tek CNC   http://www.tigerteccnc.com.cn/
- B and B Plasma   http://www.bandbsystems-inc.com/index.html

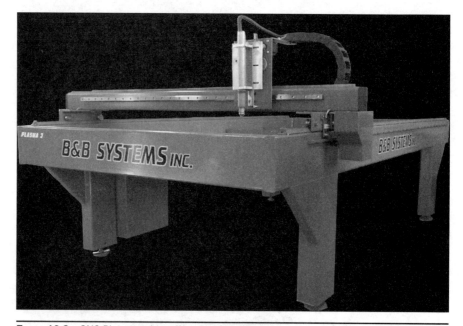

FIGURE 10-3   CNC Plasma table offered by B and B Plasma.

- K2-CNC   http://k2cnc.com/

Mills and Lathes

There is a notable difference in the construction of a mill or lathe, based on if its initial intended purpose is for CNC use. For a well-built and quality machine you will find that the CNC system design has been modified to be able to withstand the higher cutting forces and rigors that are not typical on manually operated machinery. Hence, purchasing a new or used manually operated mill or lathe with a CNC retrofit is simply not a valid comparison to the units these vendors have.

- Tormach   http://www.tormach.com/

Choosing a Ready-Made CNC System

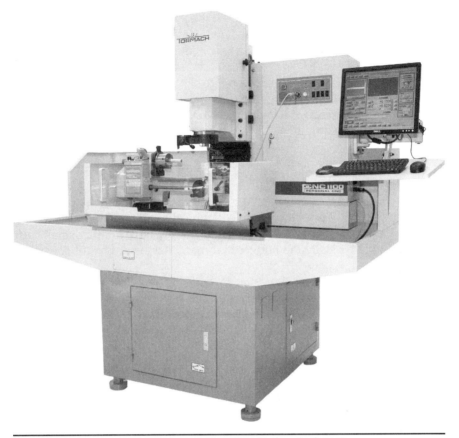

FIGURE 10-4 Tormach CNC Mill shown with optional lathe attachment.

This company engineers all of their product offerings from the ground up, and the exceptional quality is evident in their mill, Duality Lathe, tooling, etc. Everything from PTFE-bonded guide ways to using rolled ball screws are standard equipment issue on their mill. They also use the Mach3 controller software, but have developed in-house an enhanced and modified version of the Windows XP software, which is branded as MachOS. Their enhanced version of the operating system software enables the Mach3 software more stability and to operate at even higher step rates. The improvements made are transparent from a user perspective when working with the XP operating environment. If you have ever shopped for mill tooling and accessories, odds are you have encountered the Tormach tooling system (TTS). This is an invention of Tormach that allows either manually or CNC-operated milling systems quick change tooling, which can greatly improve milling operation times.

They offer fourth-axis rotary tables, but do have a Duality Lathe that works with their mill as well. If you have space constraints in your shop, this can be a lifesaver rather than having to have two pieces of dedicated equipment taking up floor space.

- MDA Precision   http://www.mdaprecision.com/

MDA Precision carries a range of small personal-sized CNC mills and lathes. They also have product offerings that are "CNC ready" so you can perform the upgrade at a later date. MDA carries the Wabeco, GOLmatic, and Prazi line of equipment and are controlled via the Mach3 controller system.

Do-It-Yourself (DIY)

Performing a simple Web search will give you many plans for DIY CNC machines. The bulk of them will typically call for wood, plastic, or aluminum for the construction materials and usually have a rather small cutting footprint. It's typically the budget that mandates what type of construction materials are used.

If you are interested in building your own table that can be used for routing, plasma, water jet, etc., then you should check out http://mechmate.com/. They offer free plans, advice, and also give recommendations where you can obtain manufactured parts if you are not able to make them yourself. They also guide the user through all of the details that are involved in building your own hardware controller and have a fairly extensive online forum for interactive questions with other builders. Using their plans you should easily be able to scale up or down a machine that suits your needs.

Vendor Listing

There are many places where you can obtain motors, drives, CNC electronics, spindles, various electronics, complete hardware controllers, connectors, raw materials, etc., for your CNC build or rebuild. Shown here is a listing of companies I have used in the past and recommend for your use:

General Electronics Parts
Marlin P. Jones   http://www.mpja.com/
Mouser   http://www.mouser.com/
Warp 9 Computers   http://warp9computers.net

Spindle Heads and VFD's
PDS Colombo   http://pdscolombo.com/
Delta Electronics   http://www.delta.com.tw/

Raw Materials
Online Metals   http://www.onlinemetals.com/

CNC Electronics
Sound Logic   http://soundlogicus.com/
Mesa Electronics   http://mesanet.com/
PMD eXpress   http://pmdx.com/
Campbell Designs   http://campbelldesigns.net/

General Hardware
Fastenal   http://www.fastenal.com/
Reid   http://www.reidsupply.com/
Enco   http://www.use-enco.com/
McMaster   http://www.mcmaster.com/
Gear Rack   http://stdsteel.com/
Bearings   http://vxb.com
MSC Direct   http://www.mscdirect.com/

Linear Guides
Modern Linear   http://modernlinear.com/
Danaher Motion   http://www.danahermotion.com/
HiWin   http://hiwin.com/
Rollon   http://www.rollon.com/
Pacific Bearing   http://www.pacific-bearing.com/

Ball/Lead Screws
Danaher Motion   http://www.danahermotion.com/
Nordex   https://nordex.com/
BSA   http://www.thomsonbsa.com/
Nook Industries   http://nookindustries.com/

Motors/Drives
Granite Devices   http://www.granitedevices.com
Applied Motion   http://www.applied-motion.com/
Keling Inc.   http://kelinginc.net/
Gecko Drives   http://www.geckodrive.com/

Publications
Digital Machinist   http://www.digitalmachinist.net/

CHAPTER 11
Building Your Own CNC Plasma Table

This chapter details on the construction process of building a plasma table capable of working with a sheet of material 4 by 4 ft. The overall dimensions of the table are approximately 5 by 5 ft, so this unit should fit comfortably in most shop environments.

The main support area for the table and legs were constructed from 3 by 3 by $3/16$-in thick wall steel tubing, with all of the joints welded (see Fig. 11-1). Now is the time to make sure the sections are square to each other. Note the pipe clamp being used to provide final adjustments prior to welding.

The spans of the four horizontal tubes are each 5 ft (or 60 inches) in length. The vertical legs are each 28 in long. The legs may appear to be rather short but when attached to the underside of the horizontal support table, along with leg levelers, the distance came out to a good working height. Note that when welding together the four members for the horizontal table, it helps if all of the lengths are the same. To help in ensuring the two upper members remain parallel to each other, I used two long sections of aluminum. The two sections were through drilled (as a pair) at both ends and were bolted to the upper horizontal members.

The leg-leveling system used here is comprised of the parts shown in Fig. 11-2. The plates are each 3 by 3 by $1/4$ with a $3/4$-in through hole bored in the center; each welded to the end of a leg member. A jam nut was first welded onto the top side of each plate, allowing a bolt with an additional jam nut to provide for some final height adjustment, but primarily for leveling the table.

Before the four horizontal members were welded together, a centerline was scribed onto each end of the top side of each of the *upper* rail members. A $1/4$-in section of steel plate 4 in in width and the same length of the members is then placed on top of each upper rail member. These plates are offset from the centerline allowing $3/4$-in overhang to the outside, which provides a mounting surface for the gear rack. Note that there ends up being a $1/4$-in overhang of the plate toward the inner edge of the members. After the four horizontal table members are welded together, the legs are welded to the underside of the upper members and to the side of the lower members (see Fig. 11-4 for a photograph of leg placement).

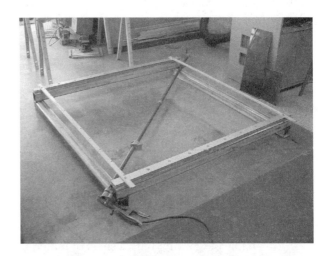

FIGURE 11-1
Materials assembled for building a CNC plasma table.

FIGURE 11-2
Leg-leveling system.

FIGURE 11-3
Complete leg bottom.

FIGURE 11-4 Squaring and attaching the four legs.

The linear guides were placed on each of the two members along the members' centerlines and holes were located for through drilling. The use of center punches for this type of operation can be very valuable. In addition, it is highly recommended that for all drilling and tapping procedures you step-drill to the final drill size. After the through holes are bored through the plate and tubing, the holes in the tubing are tapped and the guides are mounted as shown in Fig. 11-5.

After the table and leg members are welded together, the X-axis guide system rails are loosely bolted to the upper horizontal members. Final adjustment and tightening of the bolts is performed in a later stage, after which the gantry is added.

FIGURE 11-5 X-axis members with linear rails mounted. Note gear rack attached to underside of plate.

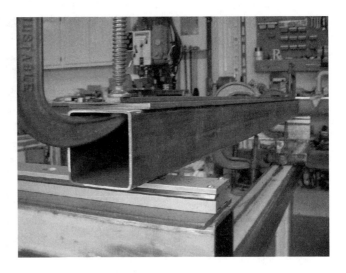

FIGURE 11-6
Addition of bar stock to act as a nut when used with the thin-wall gantry tubing.

The span of the gantry is also made using 3 by 3-in tubing, however, 1/8-in wall thickness is used. The reasoning behind using thinner wall tubing for the gantry is two-fold: it has lighter mass and also serves as an example to the reader about how thin wall tubing could be used for the entire construction of the plasma table. The difference in construction primarily involves the addition of bar-stock material to be used as a nut plate in which to bolt the guides. As 1/8-in wall tubing is too thin to obtain enough tapped threads to bolt to, a section of 1/4-in bar stock is used to provide the threading. The bar stock used is 3/4-in wide by 60 in long to match the gantry span. This is shown in Fig. 11-6. To ensure the mounting holes line up properly, the 3/4-in bar stock, 4-in wide plate, and 3 by 3-in tubing are all aligned together and through-drilled to the proper tap size (1/4-20, in this instance). Note that the bar stock stays in the centerline with the tubing and the 4-in plate gets the same side offset as in the previous guide rail to table-mounting discussion.

As was the case when mounting the X-axis rails, it is imperative that you maintain an accurate centerline when drilling the pieces for the Y-axis rail (i.e., gantry) linear guide. This helps to ensure that your X- and Y-axes stay 90 degrees or orthogonal to each other. After the plate and tubing are through-hole drilled and the bar stock is tapped, the Y-axis rail is bolted to complete the gantry span assembly.

Next, the two X-axis carriages are placed on their respective rails and each receives a 3 by 3 by 3/16 × 4 3/4-in long section of tubing that is directly bolted (from the underside) to the carriage plate. A measurement is then taken to determine the distance of the outside to outside span between the two carriages. This is the actual length of the gantry and should come out to be 62 1/4-in long. The two 4 3/4-in tubing sections are carefully aligned to the underside ends of the gantry and skip-welded to the gantry tube. Note that an

FIGURE 11-7
Carriage with 4³⁄₄-in riser tubing.

additional 4¾-in tubing section is also needed for the Y-axis carriage; hence, a total of three are needed (see Fig. 11-7).

Next, three gantry end caps are fabricated, which allow for motor-assembly mounting that drive the X- and Y-axis. These are constructed from ¼-in plate, 4-in high by 7-in wide. A dogleg is cut by removing a 3 by 4-in triangle of material, as shown in Fig. 11-8. Finally, a ⅜-in hole is bored 1½ by 2-in up from the opposite corner. This ⅜-in hole is used later on to mount and pivot the motor-reduction assembly. Once the end caps are complete, they are skip-welded in place.

It is important to note that the three end sections, where the end caps are welded, are perfectly square to the carriage plates. This ensures the pinion

FIGURE 11-8
Gantry end caps and motor-assembly pivot.

gear from the motor assembly engages the gear rack at a 90-degree angle. If off even a slight amount, both tracking and wear issues will become a problem.

The gear rack selected for this table build is 20 pitch 20 degree pressure angle. The dimensions for this particular rack are 1/2-in high by 1/2-in wide, with each section being 60 in long. The rack is bolted (via #10-24 BHCS – button head cap screws) to the undersides of the 4-in wide plates. The rack is centered in between the 3/4-in overhang with 1/8-in space between the gear rack and the 3 by 3 in tubing. This, in turn, leaves an 1/8-in gap between the other side of the rack and the edge of the 4-in plate. Holes to mount the rack are bored through the 4-in plate, centered along the 3/4-in overhang at 8-in spacing intervals. Each section of gear rack is then temporarily clamped into position in order to locate the mounting holes. Note, it is acceptable to use two or more sections of gear rack to complete a span. You must, however, ensure that the abutting rack sections are cut and ground for proper pinion gear travel.

The pulley and belt reduction units are constructed next. The particular units for this project use NEMA 23 motors with a reduction ratio of 3.5:1. Refer to Chaps. 3 and 4 that discuss motors and reduction unit types for a more detailed discussion regarding this style of reducer. Your ratio and motor footprint size may be adjusted to suit the table you are constructing. Holes, 3/8-in were drilled in these units 1 in from the corner on the rear plate. Socket head cap screws (SHCS) with 1/2-in shoulder provide the mounting and pivot action. SHCS (#1/4-20) were also threaded into the dogleg sections of the gantry end caps allowing extension springs (3/4 by 2 1/4 by 105 in) to provide the pinion gear preload. This length spring was used in conjunction with 20-tooth pinion gears. If a pinion with a different diameter is used, an alternate length spring should be chosen (see Fig. 11-9).

FIGURE 11-9
Mounted reduction unit with tensioning spring.

Figure 11-10
Mounting structure for Z-axis back plate.

Things are now at a point where the construction process, minus the Z-axis assembly, is complete and ready for primer and paint. You may either mask off the areas that do not require painting, such as the aluminum guide rails and gear rack, or disassemble everything. If you opt for disassembly, make sure you label, orient, and number each piece so things can go back together with no alignment problems. Then, wipe down all of the steel areas,

Figure 11-11
Z-axis back plate with spacer and guide installed.

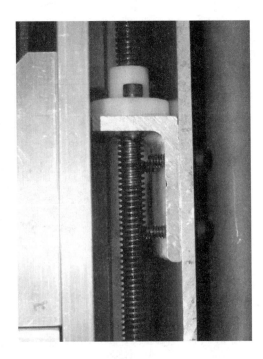

FIGURE 11-12
Angle bracket attached to lead screw nut.

FIGURE 11-13
Assembly with carriage traveler plate.

Building Your Own CNC Plasma Table 181

FIGURE 11-14
Z-axis assembly, viewed from the left side.

FIGURE 11-15
Z-axis assembly, viewed from the right side.

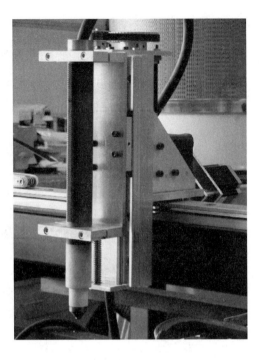

FIGURE 11-16 Z-axis assembly, viewed from the front angle.

prime, and paint. This table received a black top coat followed up with a matte clear sealer finish. Alternatively, the table could be powder coated as well.

The Z-axis back assembly utilizes aluminum plate 5 by 12 in long with the bottom edge mounted at a height equal to the underside of the gantry. Thick pieces ($\frac{1}{2}$ in) of $4\frac{1}{2}$ by $4\frac{1}{2}$-in aluminum are bolted on each side of the $4\frac{3}{4}$-in gantry tube and provide the vertical mounting of the Z-axis back plate. Take

FIGURE 11-17 X- and Y-axis complete.

care to ensure the non-dogleg corner of the triangular mounting plates are precisely 90 degrees, or else the torch position will not be exactly straight up and down to the material.

A 12-in length section of roller rail provides the linear motion. However, a 1/2-in thickness spacer is needed to end up allowing travel space for the 1 1/2-in diameter lead screw nut. In this build, the additional 1/2-in spacer yielded a 1 5/8-in gap between the face of the back plate to the front face of the rails' carriage surface.

The lower bearing mounting plate is affixed to the bottom edge of the backing plate. The plate-bearing hole is centered between the open area of the back plate (accounting for the width of the roller rail and spacer) and extends 7/8 in away from the face of the back plate. The upper mount is bolted to the top edge of the backing plate with its bearing hole criteria matching that of the bottom plate. The upper plate is, however, extended rearward to allow for mounting of a NEMA 23 motor. Using the online calculator found at http://www.sdp-si.com, a centerline distance of 2.91 in was determined for the belt and the two pulleys chosen. Note that this specific implementation is an increaser to alter the number of steps of the lead screw from 20,000 to approximately 8,000 steps/in. Once the upper and lower bearing plates are completed, they are bolted to the top and bottom edges of the back plate assembly.

Next, a 1/4-in thick carriage plate is made and bolted to the carriage traveler. Careful measurement is taken between the back of the carriage plate to the centerline of the lead screw assembly. Using some 1/4-in aluminum angle, a bracket is made that rides between the back side of the carriage plate and the underside of the lead screw nut. Four bolts threaded into the angle bracket hold it to the back side of the carriage plate with three bolts attached to the nut. It is very important that the nut's angle bracket line up squarely to the back

FIGURE 11-18
Completed table.

FIGURE 11-19
Completed table.

side of the carriage plate such that no flexing or binding occurs. If needed, shim or provide the necessary adjustment until there is an exact fit.

The machine torch is mounted to the carriage traveler plate via two clam-shell mounts. Note that the length of the carriage plate was determined by where the mounts hold the torch at its widest point. If you do not have a machine torch for your plasma cutter, the hand-held torch can be used. You will obviously need to alter the mounting schema as the clam-shell method will not work.

All that is required at this point to complete the table assembly is motor wiring, safety covers for the reduction units, the mounting of optional limit/homing micro switches, and a grid surface for the table (not shown). There are several methods in which to provide the grid and the spacing is generally determined by the overall size of the smallest parts the user will be cutting. Obviously a tighter spacing grid is needed in order to support smaller pieces. One approach to constructing a grid is to use, for example, $1/8$-in thickness mild steel $1\frac{1}{2}$-in wide standing on the edge every 2 in apart from each other. These pieces are held on their edge by spacers tack welded to the table with gaps in between them. The user may opt to half-lap the sections of material. This latter approach is labor intensive and the reader should realize the grid material will require replacement after several hours of cutting time. An easier alternative is to use 1-in angle iron placed on the table structure with the pointed end of the angle facing upward. Be advised that small gaps are needed between the angle iron segments for proper cutting operation. Figures 11-18 and 11-19 show views of the completed plasma table.

PART V
Appendices

APPENDIX A
Project Implementation and Examples

Examples of Items that Can Be Produced on a CNC Router

As people are often new to the capabilities of what can be produced on a CNC machine, they typically inquire as to what can be achieved. Included here (and throughout various applicable sections within the book) are examples of products that the author has made. Included are examples of both 2D and 3D items, as well as a simple example of a fourth-axis rotary turning. All of the examples listed are only a sampling of what can be produced on a CNC router that range from woodworking tools, various signage, and building musical instruments. The limitations are typically the user's imagination, material type with hold-down method, and programming techniques.

Monument Sign

The first of the examples is a monument sign (Fig. A-1). The entire structure (posts and panel) are comprised of an outdoor, weatherproof composite-sheet material, sized and cut on a 4 ft × 8 ft CNC router table. The raised panel effect on the posts was done using a flat-bottomed router bit with ogee side profiles. The finials are also made of the same material and were laminated into a thick block. The block was then turned on a rotary fourth axis in lathe mode to produce the desired result. The wide cap over the panel is multiple segments of the same identical size and shape – all laminated together to achieve the needed depth (width). The lettering is made of PVC (polyvinyl chloride) sheet material and is glued on the panel substrate. Although this is a rather large and involved project, all of the parts are comprised of sheet stock and all routing is simple 2D.

Furniture

One of the first things produced on the author's CNC machine were shop cabinets (see Fig. A-2). These particular cabinets were 36 in wide, 31 in high, and 12 in deep. The sizing of this cabinet is specific, as one single 4 ft × 8 ft sheet of material will yield one cabinet including the interior shelving. These were constructed from $3/4$-in MDF, and rabbet joinery was used throughout the construction for both assembly registration and stronger glue joints.

Figure A-1 Monument sign.

Tools and Templates

CNC machines are often used by woodworking and metalworking shops to produce patterns or templates that are used in the construction of the final item being produced. Other times, as in the following example, a CNC router is used to produce various parts of tools used in the shop. Shown here are three different types of dovetail jigs (half-blind, variable spaced through dovetail, and fixed dovetail) that are used in conjunction with a hand-held router to

Figure A-2 Shop cabinetry.

FIGURE A-3 Dovetail jig #1.

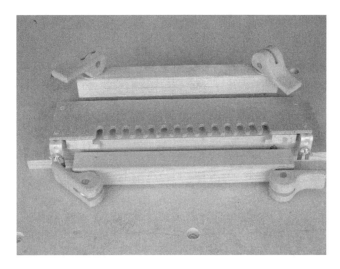

produce the dovetail joints. The sample pieces shown with the dovetail joints cut were made with fixed-space dovetail jig templates shown in Fig. A-3. Note the gap in the joint is intentional to show clarity of the cut pieces (see Figs. A-4–6.)

In another project done by the author, a specific tool was required to bend a specified radius on some fret board wire for a guitar neck. Although tools, such as this one, are readily available from luthier supply outlets, with a CNC router, construction of tools becomes fairly easy and can be built and ready for use sooner than going to the store or waiting for shipping. By building your own tools you can often times save a great deal of money plus end up

FIGURE A-4 Dovetail jig #2.

FIGURE A-5 Dovetail jig #3.

with a better tool than what can be purchased. Figure A-7 shows a fret wire bending tool constructed from $1/2$-in clear acrylic. Using the CNC machine you can also produce various types of patterns for use with other tools, such as using a hand-held router. There is money-making potential for producing high-quality precision patterns, as not everyone will own or have access to a CNC routing machine.

ADA Signage

A specific type of signage that is typically produced on a CNC engraving machine is referred to as "ADA signage," where ADA stands for American Disabilities Act. The following two examples were made by the author using

FIGURE A-6 Sample of DT joint cut.

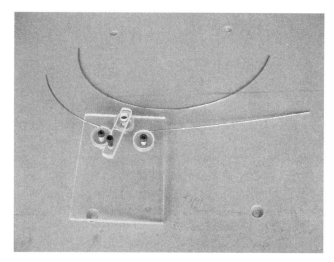

FIGURE A-7 Fret wire bending tool.

matte and frosted acrylic with colored appliqué applied, routed, and weeded away to leave the resulting lettering, numbers, and logo (see Fig. A-8). In addition, the CNC router/engraver is used in the process of Braille lettering shown in both examples.

There are three main methods in which to produce Braille. One method is to use dye casting and pour material (liquid plastic) into the mold that has recessed areas to fill the voided areas. The second popular method is to

FIGURE A-8 ADA restroom sign.

Figure A-9 ADA sign.

use a hollow-point bit, specific to the signage industry, that when plunged to a specified depth in the material leaves a raised bump. A second area-clear process is then required to remove the remaining material from around the individual Braille letters. The third method, and the one used by the author, is to employ a specific sized end mill to drill holes into the ADA material stock at specified locations. Then, using either a hand-held or CNC-held tool, small raster beads (typically made of glass or metal) are injected into the holes, leaving the raised areas.

V Carving

An elegant looking type of incised lettering is most commonly referred to as V carving. This is accomplished using a router bit with the cutting tip forming a "V". V-router bits are produced in varying angles (discussed in more detail in the tooling section of the book); both of the examples shown here were made using a 90-degree bit angle cutter. The first example shown has lettering, a border, and accents carved into a piece of $3/4$-in northern ash. The wood was then face-painted black and passed through a surface planer, leaving the painted engraved areas. The second sample material is $1/2$-in clear acrylic. The back of the acrylic substrate was first painted and then reverse engraved (i.e., from the back). Once the "S" was engraved, a second paint color was applied to the back of the acrylic, filling in the engraved area. As the back of this sign was already painted there was no need to do any masking for the fill color (see Fig. A-11).

FIGURE A-10
V-carved house sign.

Bevel-Edge Lettering

For simple geometries containing all "outer" corners (no interior sharp corners) any software method (CAD, CAM, or manual) can be used to produce items with a beveled edge. To achieve the full effect that bevel-edge carving has to offer (such as interior corners and height variations), you will require software that supports this feature. As is the case in the V-carving process, the wider the stroke of the font area, the deeper the bit will cut into the material. Similar to this, bevel lettering also varies in height, based upon the width of the stroke of the font being used. Note the changes in elevation of

FIGURE A-11
Lettering reverse-engraved into 1/2-in acrylic.

FIGURE A-12 Bevel edge letter example.

the ampersand sample shown in Fig. A-12, where the wider font strokes are taller in height than the narrower sections. The particular material used here is painted PVC.

Raised/Recessed-Lettering Effects

This example shows a commonly produced type of sign that employs both raised- and recessed-lettering effects. The blank of material starts out as one flat-faced block of material and the lettering, border, and logo are effectively raised as the area around them (the background shown in black) is removed using an area-clear process. The lettering of the company name at the top of the sign is area-cleared into the material blank yielding a recessed-lettering effect. This particular sign is made from high-density urethane (HDU) foam and was primed and painted in one color prior to the routing process. Prior to routing, the finished blank was covered with a sheet of masking, allowing only the routed areas to be painted a contrasting color after the area-clear process was complete. A light texture was carved into the background to simulate a light sandblasting effect.

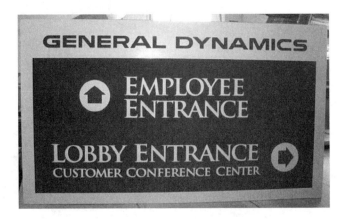

FIGURE A-13 Raised/recessed artwork in HDU foam.

FIGURE A-14 Wood grain pattern in MDF.

Textured Background Effect

It is often desirable to work with a composite material and apply a simulated texturing effect to it rather than sandblasting or relying on time and the elements to produce a weathered effect. In the image shown in Fig. A-14, a simulated wood grain pattern was 3D-carved (i.e., CNC routed) into a piece of medium density fiberboard (MDF). This capability allows the CNC user to work with varying materials that may not intrinsically already have a grain and can easily simulate a sandblasted or other types of design. CAM software will often have varying samples of simulated textures. However, by scanning or modeling software, virtually anything can be programmed for dimensional routing as a background texture. Other common examples include diamond plate, pebble, stone or granite, rope, hammered or chiseled, and weaves; the possibilities are unlimited.

Filled-Shallow Engraving

With some imagination, various finishing techniques can be applied to standard routing to yield an equally dramatic effect. In the piece shown in Fig. A-15, the composite material was area-clear engraved with a fine-tipped conical routing bit to a finished depth of 0.040 in. The routed area was then filled with a mixture of two-part epoxy that was mixed with orange paint. Once dried, the surface was sanded flush to the substrate with a fine-grit sandpaper. The end result yields a shiny inlay effect.

Accented Fonts

There are a great number of TrueType Fonts (they all have a .ttf extension) installed on most computer systems. In addition, there are literally thousands of fancy aftermarket fonts that are readily available to the user either free via Internet downloads or as libraries that can be purchased. In the sample shown in Fig. A-16, an accented font was used to "spice up" the lettering. This particular example is constructed of a PVC base material with textured engraving stock laminated to the surface (prior to the routing/engraving

Figure A-15
Epoxy-filled engraving.

process). A "V"-tipped cutter is first used to route the lettering accents using a 3D carving routine. Once complete, an end-mill style of cutter is used to profile-route the interior and exterior boundaries. This same approach can be used in wood or virtually any material to also give a shadow-type effect.

Pulley Blocks

A few years back, the author was subcontracted to manufacture the components needed to reflect the look of pulley blocks that were typically used

Figure A-16
Accented letter made from engraving stock laminated to PVC base.

FIGURE A-17 An exploded view of a pulley block.

on sailing ships hundred of years ago (Figs. A-17, A-18, and A-19). A total of five different sized blocks were manufactured for configurations of both single- and double-sheave pulleys. After assembly, these pulley blocks were distressed and dipped in boiled linseed oil and were used as working units for the Walt Disney "Pirates of the Caribbean" movies. The author has produced several of these since that were custom engraved and given as gifts to various nautical enthusiasts. Shown in Fig. A-20 is an assembled and working version of a pulley.

FIGURE A-18 Pulley block showing the components in both exploded and loosely assembled views

Figure A-19
Sheaves for the pulley blocks. Various sized sheaves that were turned on a fourth axis to cut out the rope groove along the edge of the acetyl plastic.

Oversized Projects

The CNC user may often be required to produce items that are in excess of the working envelope of the machine. This was the case in a project the author did for a local Dallas, Texas, museum exhibit (see Fig. A-21). This representation of the United States is comprised of HDU foam. As the specifications for this called for the project to be overall 8 ft in height and 15 ft in width, it was constructed in four segments (note the working envelope of the CNC router used was 4 ft × 8 ft). The individual portions were then glued together to make one large continuous panel. The periphery of the panel was built up using five additional courses of the same 1-in thick HDU material to a final depth of 6 in.

Additional exhibit items manufactured were numerous sized domes (also made from HDU foam) in the shape of campaign buttons (see Fig. A-22).

Figure A-20
Assembled and working pulley block.

FIGURE A-21 USA panel.

Backlit Acrylic LED Sign

Very striking results can be achieved by the use of light emitting diode (LED) backlighting when used with either clear or colored acrylic. In the example shown in Fig. A-23, a sheet of ½-in thick clear acrylic was first painted black on one side. This same side was then "reverse-engraved" and portions area cleared to allow light from the various colored LED lights to illuminate the carved areas. As desired, all same colored LED lights (such as white) can be used with color changes, achieved with the use of transparent colored paints, in the carved accent areas. Here a transparent red paint was used to highlight this company's tag line.

FIGURE A-22 Campaign buttons.

FIGURE A-23 LED backlit sign.

3D Routing

The user often will need to produce items that necessitate the CNC machine to simultaneously operate all three (i.e., X, Y, and Z) axes. This is commonly referred to as 3D routing; several examples are included here. Prior to the finish routing pass it is usually necessary to hog-out the bulk of material that needs to be removed. It is most advantageous to use a larger sized end mill (or a series of end mills) in a series of roughing passes to leave only a small amount of remaining material for the finish pass, as this saves a great deal of machine run time. The final pass usually is performed with a small ball-nose router bit. However, any size or style of bit can be used in conjunction with rather small step-over passes to achieve a smooth finish. Note that the final or series of final bits should correspond to the geometry of the item you are making. If your product has steep vertical angles in its geometry, the use of angled V or conical bits should be avoided, as the side of the angled bit will limit the vertical angles.

Shown in Fig. A-24 is a 3D rosette carved into a 4 in × 4 in section of pine. As no steep angles are present, a 30-degree -0.010-in flat-tipped conical engraving bit was used with a step-over of 0.0025 in. The smaller the step-over, the smoother the surfaces; however, the time required to finish the carving increases. Experience with both bit types and material choices will yield varying results that the user needs to explore for themselves.

FIGURE A-24   3D rosette.

The second example (see Figs. A-25 and A-26) are one of a series of exterior monument signs (shown here of unpainted HDU foam) for a well-known electronics company's headquarters located in Fort Worth, Texas. The largest of these was made from a blank 3-ft square and was 8 inches thick. An $1/8$-in ball-nosed bit was used in the final pass, which took several hours to complete.

The final example is of a customer design, which was carved out of builder's insulation foam (found at local building supply outlets; see

FIGURE A-25   Logo top.

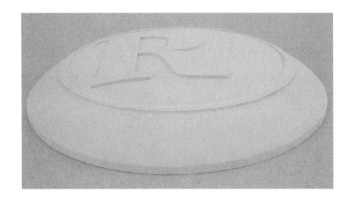

FIGURE A-26 Logo, side view.

Fig. A-27). Low-cost foam was used here first for customer approval prior to the final carving in an expensive exotic hardwood.

Panel with Removed Lettering

The panel signs shown in Fig. A-28 are made from $1/8$-in aluminum sheets with the logos and lettering removed. The aluminum panels are powder coated (or

FIGURE A-27 Horse carved in insulation foam board.

FIGURE A-28
Marquee sign installed at a local outlet mall.

painted) and then backed with $1/2$-in thick colored acrylic and installed in this marquee that has lighting installed behind the signage panels. The lighting provides nighttime illumination of the cutout areas.

Engraving Using Engraving Stock

Engraving stock is various types of plastic (typically acrylic face with ABS plastic core), are available for both laser-specific and rotary engraving and range from $1/32$ in up to $1/4$-in or more in thickness. It is available in almost any color (face and engraved color), texture, and surface finish (satin, matte, shiny lacquer, brushed metal, wood grain, and granite). Full sheets typically come 2 ft × 4 ft; however, are available as $1/2$, $1/4$, and $1/3$ sheets as well. Although any type of rotary cutting bit can be used, fine conical-tipped bits are preferred – especially for detailed types of work. The most common stock is generally $1/8$-in in thickness and is two-tone in color. The engraving examples shown in Figs. A-29 and A-30 are $3/32$ in thick black material, with an engraving depth at 0.012 in for both. Note this type of engraving is usually performed at a specified or suggested depth by the manufacturer.

The first graphic (Fig. A-29) shows an example of what is commonly referred to as "stick font" engraving. This type of font is different from TrueType style in that each stroke of the font is a single vector entity. Hence, the width of the stroke on the font is directly dependent upon the diameter of the rotary cutting bit being used.

Shown in Fig. A-30 is an example of area-clear engraving using a standard TrueType font. The bit used for this is a flat-tipped conical cutter with a 0.005-in cutting width.

FIGURE A-29 Stick font engraving.

Musical Instrument – Solid-Body Electric Guitar

Luthier shops can benefit from the use of a CNC routing machine for numerous aspects in the construction of musical instruments. Tasks such as inlay, building forms for steam bending, pocket hole routing for pickups and tremolo systems, and body and neck shaping are but a few common tasks.

Shown here in various detailed steps is a project the author recently completed. Please note that almost this entire project was performed on a shop-built CNC router with only light sanding required to the finish the final product.

Obviously, a finished guitar such as this or the various components to assemble your own instrument can be purchased, but you will find it much

FIGURE A-30 TrueType font engraving.

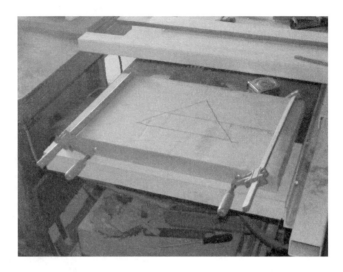

FIGURE A-31 Gluing up the edge jointed body blank.

more rewarding to custom make your own from scratch – not to mention much less expensive. This guitar is patterned after a FenderStratocaster© that is outfitted with a 1970's style neck. All of the CAD drawings were produced by the author who lofted dimensions from various other guitars. As this project incorporates many cutting files for both 2D and 3D routing, uses several differing bit sizes and styles, a semi-detailed overview of the construction process is covered here. If the reader has a comparable project in mind to build, just remember to tackle each step, one at a time, paying particular attention to hold-down methods as applicable. You will ultimately end up with a final project you can be proud of.

The body of this instrument started out as an 8/4-in rough-sawn timber of alder. Two halves were edge jointed and glued together and the router was used to dimensionally thickness the blank by routing from each side (see Figs. A-31 and A-32). Almost all of the required routing on the face and back of the body is straight forward 2.5D routing (pocket routing to varying depths). Three-dimensional routing was required for the flat-plane forearm bevel on the face of the body (see Fig. A-33).

The body was then flipped over to perform routing operations on the back that include: pocket for spring and tremolo cavity, as well as the contoured tummy cut. Note the Z level roughing passes in Fig. A-34 that take place prior to the finish pass to save cutting time. The finished contour bevel can be seen in Fig. A-35. To further illustrate the effect of the two 3D-cutting operations, a side shot of the guitar body is shown in Fig. A-36. Note the parallel angles of the forearm angle and tummy-cut contour.

The round-over along the perimeter on both the face and back were accomplished using a hand-held router with a round-over bit. An important realization that the reader needs to understand is that not all operations are fastest or best suited to be performed on a CNC machine. Assuming you

Figure A-32 Using CNC router, the dimensional thickness of the blank.

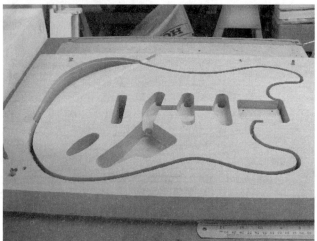

Figure A-33 Face shot of guitar body.

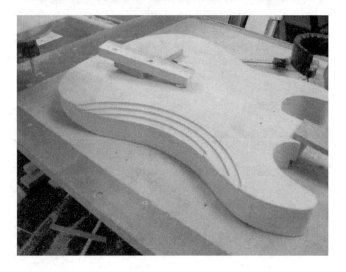

Figure A-34 *Z* level roughing passes for the contour tummy-cut.

FIGURE A-35 Side of the completed guitar body back.

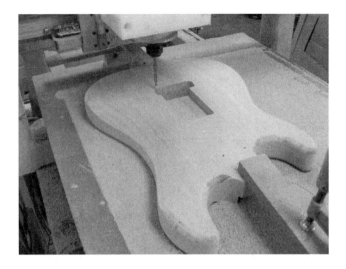

have access to a hand-held router and the appropriate bit (in this example), the operation is actually faster and just as accurate to do manually. The author routinely (as applicable) uses various manually operated tools, such as a hand router, table saw, band saw, etc. in the production of many items. Both common sense and experience play a critical role in the CNC user's approach to manufacturing.

Note that in cases where routing is to be performed on both the front side and back side of material and the material faces do not sit flat, it is advisable to build some type of a holding jig (such as that shown in Fig. A-37). This allows the user to solidly hold the stock during cutting operations as well

FIGURE A-36 Side elevation showing 3D contouring cuts.

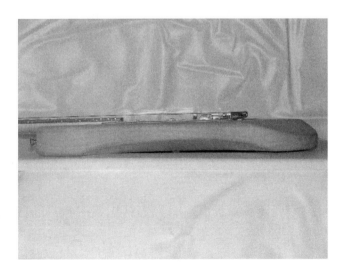

FIGURE A-37 Neck holding jig.

as realigning the the material on the spoil board. This was the case in the construction of the neck for this guitar.

The neck is made from hard maple and it has a tapered oval shape applied to the back side of the neck (see Fig. A-38). The taper along the length is 0.050 in thicker toward the base of the neck. Once the oval shape and transitions (Fig. A-39) were complete, the neck was flipped over and a channel was routed to house the dual-action truss rod (see Fig. A-40).

The fret board shape was perimeter routed to a depth of $5/16$ in into a 4/4 in blank of rosewood and the fret board blank removed by resawing to

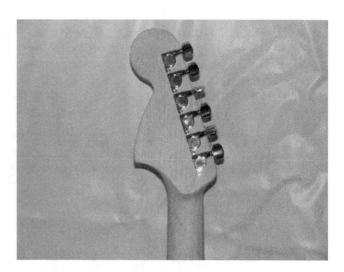

FIGURE A-38 Tapered oval on back of guitar neck.

FIGURE A-39
Transition to head stock from neck back side.

a thickness of $^5/_{16}$ in on the table saw, as seen in Fig. A-41. This photograph shows two rosewood fret board blanks ready to be resawn on the table saw (one in preparation for building another guitar neck).

The fret board was then carefully glued to the face of the maple neck, covering the truss rod. The next step was to thickness the rosewood blank to a constant thickness of $^1/_4$ in as shown in Fig. A-42. Once the fret board was thickness-dimensioned, work began creating the fret board radius. Note that in lieu of a constant radius fret board (which is typical), this one received a compound radius that starts out as a 10-in radius at the head of the neck and

FIGURE A-40
Channel with installed truss rod.

FIGURE A-41 Two rosewood fret board blanks ready for resawing on the tablesaw.

ends in a 16-in radius at the heel of the neck (see Fig. A-43). Then, using a 0.020-in end mill, the fret slots were routed to the proper depth in a series of passes (Fig. A-44). Note that by using a CNC router for this operation, the spacing of the fret slots relative to the nut can be maintained to high tolerances. This is critical in the individual notes when playing the instrument.

Final work on the neck was to thickness the headstock (from the face side) to $^9/_{16}$-in thick, route the holes for the tuner mechanisms, and, finally, add a transition radius between the fret board and the head of the neck (see Fig. A-45). Figure A-46 shows the fret slots and completed inlay markers. It also shows the completed fret slots (with loose fitted fret wire) along with

FIGURE A-42 Routing fret board to proper thickness.

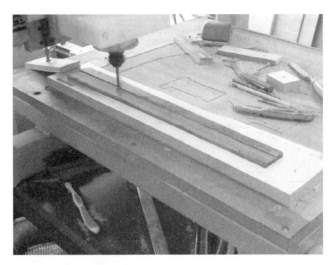

FIGURE A-43 3D routing of a compound radius on the fret board.

FIGURE A-44 Partially cut fret slots.

FIGURE A-45 Thicknessing the neck headstock.

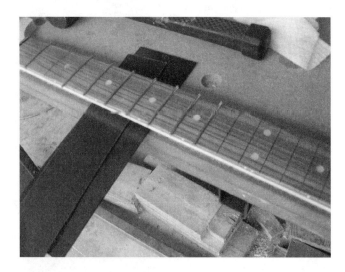

FIGURE A-46 Fret slots and completed inlay markers.

the installed pearloid fret board markers. Installing the fret wire and the remaining touches did not require any CNC work. The end result of the project is shown in Fig. A-47.

Process of Making Stud-Mounted Lettering and Logos

One of the tasks sign makers often need to perform is to produce stud-mounted letters. For those who are unfamiliar with what stud mounting is, it involves the mounting of threaded studs in proper locations (matching the same locations on a mounting pattern) on the back side of each letter or logo; generally each letter receives multiple mounting studs. During the installation procedure, corresponding holes are drilled in the wall surface to accept the protruding studs. I have gone through the process of how to perform this function countless number of times for fellow sign makers. If the studs are not done on a CNC router, the lettering needs to be manually marked and drilled by hand by the sign shop worker, which is a laborious process.

To start the process off, you will obviously need to have the vectorized artwork ready to work with, as is shown in this example in Fig. A-48.

At this point you use your CAD program to apply circles that correspond to the diameter of the mounting studs you will be using. As you can see, there are a series of stud holes placed on each of the letters in Fig. A-49. Dependent upon the material type and thickness, along with the overall height of each independent graphic, you will need to determine how many or how few studs are required to safely adhere the lettering to the wall.

It is at the following locations, shown in Fig. A-49, that a mounting pattern (typically pen plotted on a vinyl cutter) should be produced. This paper pattern is ultimately held up to the wall and the stud-mounting locations marked and drilled.

FIGURE A-47 Completed guitar displayed on stand.

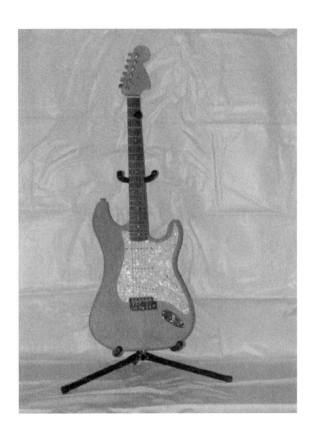

FIGURE A-48 Vector artwork.

FIGURE A-49 Vector artwork with the addition of stud-hole locations.

Figure A-50
Mirrored artwork with stud-hole locations.

The next step in the process is to mirror the artwork with the stud holes included. The intent of this step is to make sure you are viewing (and ultimately working with) the back side of the lettering to be cut. If the material you are working with has a "good side" to it, you will need to make sure that it is placed good-side face down when you start the cutting process. A picture showing the mirrored artwork is given in Fig. A-50.

This next step is not required; however, if you do have software that performs a nesting function, I highly recommend it as it can greatly save on the overall amount of material to be used. The distance between the lettering cannot be closer than the diameter of the profiling bit you will be using, but it can be advantageous to leave some remaining material for rigidity while being machined. For clarity purposes, I have exaggerated this gap distance, as is shown in Fig. A-51.

Now you are able to use your CNC router to issue a drilling routine on the stud-hole locations. Depending on your CAM software, you have the option of drilling stud holes for individual letters and performing a subsequent profiling. I find it easier to drill all of the stud holes first. Pay particular attention to the depth of the mounting holes as the objective is to create blind holes (i.e., not going all the way through the material). Obviously the holes will need to be deep enough for a flat-bottom tap to create a minimum of a few threads for the studs to screw into. Also, by drilling all of the stud holes independently from profiling the letters, you can use different sized bits for each process. If you have properly followed this procedure, you will end up with lettering that has stud-hole locations that precisely match those on the paper pattern, without the need to manually hand drill the holes.

Relay Cover Removal Job

In this example, the router is not being used to directly produce an item, but rather to *modify* an electronics part – a 12-V printed circuit board-mounted relay. Here the protective cover over the electronics portion needed to be removed from a total of 5,000 units. This ultimately freed up some valuable real estate space on some overcrowded electronics boards. A picture of the

FIGURE **A-51** Nested vector artwork.

relay as supplied is shown in Fig. A-52. The task at hand was to route the bottom perimeter of the relay using an $1/_8$-in end-mill bit, cutting through the glue bond that was holding the cover to the base.

After considering several methods in which to hold the individual relays for cutting, it was determined that the best method was to construct a tray to tightly hold 100 of the relays at one time (see Fig. A-53).

Each of the cutouts in the tray house the relays, which was carefully machined to tolerance such that a slight friction fit would occur. In addition, the depth of each socket allowed the parts to "bottom out," thus providing a consistent height of the parts throughout the tray (see Fig. A-54).

A CAM program-generated cutting file was produced such that an $1/_8$-in end-mill cutting bit would traverse the perimeter of the relay, cutting 0.030 in of the material at a depth of 0.025 in. A view of the postcut operation can be seen in Fig. A-55.

After each row of relays was routed, the units were removed from the tray (while the machine was still cutting subsequent rows) so that the operation was not interrupted. When the G-code file did reach the end, an M47

Figure A-52 Twelve-volt relay.

command would rewind the program and restart from the beginning of the file. The resulting cut relay is shown in Fig. A-56.

Unlimited Possibilities

There are literally an unlimited number of possibilities of uses for a CNC machine. Additional examples of tasks the author has produced are molds, stamp and dye pairs used in metal forming, and embossing, along with a wide range of delicate inlays for various customer projects. Hopefully these routing examples will somewhat reflect similar operations in the production of

Figure A-53 Relay holding tray.

Project Implementation and Examples 219

FIGURE A-54 Uncut relay in tray socket.

your product. Hopefully, they have shown the reader the diversity of cutting operations that are possible.

Programming Examples

There are many ways to approach program generation and execution. One way is to solely rely on the output generated from the postprocessor file used by your CAM. Yet another way is to manually write each line of code yourself – with or without the aide of a G-code programming editor. A third method is to use a combination of the previous two approaches.

FIGURE A-55 Cut relay in tray socket.

FIGURE A-56 Relay cover and internals.

I tend to favor the third method of producing cut files, especially when there are multiples of the same part needing to be done. There are several reasons behind using a combined approach:

- You are only dealing with one instance of G code for the artwork rather than having multiple instances of the same file to work with. This makes editing and file-generation times much shorter.
- As only one instance of the corresponding cutting codes is in a cut file, the resulting size of the file to process is greatly minimized.
- It is an easy and straightforward edit to either add or subtract cutting locations within the working envelope. You also have the ability to easily override cut locations you may wish to avoid or easily single out one particular location for testing purposes.
- It is an easily customizable way to control the order and direction that the cutting operations will take. You can mandate the cut order go left to right on one row and then work backward from right to left on the next row, etc.
- The overall time it takes to execute the cutting operation can be minimized through the manual control in the order of operations as well.

For those individuals who wish to take CNC programming to the next level using macros, I recommend a comprehensive text on parametric programming entitled "CNC Programming Using Fanuc Custom Macro B" by S. K. Sinha.

Example 1

We start off with a simple example of a *main* file calling a *subroutine* that contains some blocks of G code that are either manually written or CAM generated. The purpose of this simple example is to illustrate to the reader the concept and implementation of how to use subroutines. Note that throughout each of the example files, comments on what the block of code performs will be denoted as applicable alongside of each block of code.

```
**Main Program Start**
N1      G90 G20 G49      (initialization string)
N2      M03 S18000       (turn on the spindle head @ 18,000 rpm)
N3      G0 X0 Y0         (rapid move to the home location of 0,0)
N4      M98 P1000        (call subroutine number 1000)
N5      M05              (turn off the spindle head)
N6      M30              (program rewind and end)
**Main Program End**


**Subroutine Start**
O1000    (subroutine label 1000. Note that "O" is a letter, not number zero)
*Start* of cutting/engraving blocks of G-Code
  ...           (this area is where the programmer would add either manually)
  ...           (written or CAM generated lines of G code)
  ...
  ...
  ...
*End* of cutting/engraving blocks of G code
M99      (exit subroutine and return to next line of code in the main)
**Subroutine End**
```

Let's take a closer look at example 1 to discuss what is happening. The program starts out with an initialization string on line N1 to make sure that all moves and modes are properly set. G90, G20, and G49, respectively, instruct the modes to use absolute, inches, and cancel any tool length offsets that may be present. Line N2 issues the command to start the spindle head turning in a clockwise direction at a rate of 18,000 rpm's. Line number 3 is simply a command to "rapid" quickly to the home location of $X = 0$ $Y = 0$. Line N4 uses the M98 command to call subroutine labeled O1000. Hence, now the focus of processing is in the subroutine and stays there until it reaches the M99 command, indicating the end of the subroutine. Processing focus then returns to the main, but at the next line (N5), which then disables the spindle head and M30 on line N6 instructs the program to rewind to the beginning and end.

Example #2

We now delve a bit deeper with the same file used as in example 1, however, we will now use the *main* file that calls the same *subroutine* for execution in four differing temporary offset locations on the CNC work table.

Main Program

N1	G90 G20 G49	(this is an initialization string)
N2	M03 S18000	(turn on the spindle head)
N3	G0 X0 Y0	(rapid move to the G54 home location of 0,0)
N4	G52 X0 Y0	(first temporary part location that also corresponds to G54 0,0 location)
N5	M98 P1000	(call subroutine number 1000)
N6	G52 X5 Y0	(second temporary part location at new location)
N7	M98 P1000	(call subroutine number 1000)
N8	G52 X5 Y5	(third temporary part location at new location)
N9	M98 P1000	(call subroutine number 1000)
N10	G52 X0 Y5	(fourth temporary part location at new location)
N11	M98 P1000	(call subroutine number 1000)
N12	G52 X0 Y0	(cancel using temporary locations)
N13	M05	(turn off the spindle head)
N14	M30	(program rewind and end)

Subroutine

O1000   (subroutine label 1000. Note the "O" is a letter, not number zero)
Start of cutting/engraving blocks of G code
...   (this area is where the programmer would add either manually)
...   (written or from CAM the lines of G code that get used in each)
...   (instance at each temporary part location)
...
...
...
End of cutting/engraving blocks of G code
M99   (exit subroutine and return to next line of code in the main)

Let's take a closer look at example 2 to discuss what is happening. Again, example 2 is similar to the first example, however, it has some G52 commands inserted allowing temporary offset locations. Temporary offset locations allow the same execution of code to be performed at differing locations within the working envelope.

Line number 3 is simply a command to "rapid" quickly to the location that is set at G54 X0 and Y0 for preparation of getting into a "temporary offset" mode (i.e., G52). Then, on N4, we are now in the temporary offset mode at location X0 and Y0. This is the first "temporary part location." N5 then calls a subroutine with the label O1000. Note P is the subroutine calling identifier and the letter O denotes the subroutines label. It is the numeric value associated with the P and O that allow multiple subroutines to be called (such as O1001, O1002, etc.). Once the M98 P1000 command is encountered, the program focus leaves the main routine and goes to the O1000 subroutine and continues to process whatever code is contained within. Once the M99 command is reached in the subroutine, the focus exits the subroutine and returns focus to the next line in the main routine. In this case, N6 is the next line to be executed.

In short, whatever is contained in subroutine O1000 will be performed at X0 Y0, then at X5 Y0, then X5 Y5, and, finally, X0 Y5, making a counterclockwise loop starting from the lower left corner. When all of the locations have been addressed, line N12 commands a G52 at location X0 Y0 to clear the logic of any temporary work offsets (i.e., back in G54 mode). Line N13 turns off the spindle head and the final line containing M30 issues a rewind and end of the program that was running. Also, of importance, is to note that subroutines can either be part of the same file that the main is in, or can be a different file altogether. In example 1, the subroutine is shown as being included in the same file as the main routine. Then, too, it is possible to call another subroutine from within an existing subroutine. Typical nesting of this type generally extends to four levels of logic.

APPENDIX B
Programming Examples in G Code

There are many ways to approach G-code program generation. One way is to solely rely on the output generated from the postprocessor file your CAM program uses. Yet another way is to manually write each line of code yourself – with or without the aid of a G-code programming editor. A third method is to use a combination of the two previous approaches.

I often tend to favor the third method of producing cut files, especially when there are multiples of the same part needing to be done on a sheet of material. There are several advantages behind using this type of a combined approach:

- You are only dealing with one instance of G code for the artwork rather than having multiple instances of the same file to work with. This makes editing and file-generation times much shorter.
- As only one instance of the corresponding cutting codes is in cut file, the resulting size of the file to process is greatly minimized.
- It is an easy and straightforward edit to either add or subtract cutting locations within the sheet of material you are working with. You also have the ability to easily override cut locations you may wish to avoid, such as a defect or void in the material, or easily single out one particular location for testing purposes.
- It is an easily customizable way to control the order and direction that the cutting operations will take. You can mandate the cut order go left to right on one row and then work backward from right to left on the next row, etc.
- The overall time it takes to execute the cutting operation can be minimized through the manual control in the order of operations as well.

Shown here are two examples of G code that make use of *subroutines*. For various repetitive tasks, I routinely will write a *main* G-code program that determines the location(s) of where on the sheet material a common piece of G code is to be repeated multiple times. The contents of the subroutine are something the user can manually write, or, as I typically do, use a CAM program to generate the code and paste it into a separate subroutine file, which is called by the program's main file.

For those individuals who wish to take CNC programming to the next level using macros, there is a comprehensive text on parametric programming entitled "CNC Programming Using Fanuc Custom Macro B" by S. K. Sinha, that I highly recommend.

Example 1

We start off with a simple example of a *main* file calling a *subroutine* that contains some blocks of G code that are either manually written or CAM generated. The purpose of this simple example is to illustrate to the reader the concept and implementation of how to use subroutines. (Note that throughout each of the example files, comments on what the block of code performs will be denoted as applicable alongside of each block of code.)

```
**Main Program Start**
N1      G90 G20 G49         (initialization string)
N2      M03 S18000          (turn on the spindle head @ 18,000 rpm)
N3      G0 X0 Y0            (rapid move to the home location of 0,0)
N4      M98 P1000           (call subroutine number 1000)
N5      M05                 (turn off the spindle head)
N6      M30                 (program rewind and end)
**Main Program End**

**Subroutine Start**
O1000   (subroutine label 1000. Note the "O" is a letter, not number zero)
*Start* of cutting/engraving blocks of G code
...             (this area is where the programmer would add either manually)
...             (written or CAM-generated lines of G code)
...
...
...
*End* of cutting/engraving blocks of G code
M99     (exit subroutine and return to next line of code in the main)
**Subroutine End**
```

Let's take a closer look at example 1 to discuss what is happening. The program starts out with an initialization string on line N1 to make sure that all moves and modes are set properly. G90, G20, and G49, respectively, instruct the modes to use absolute, inches, and cancel any tool length offsets that may be present. Line N2 issues the command to start the spindle head turning in a clockwise direction at a rate of 18,000 rpm's. (Note that this is true if an actual spindle head is being used. If your machine has a router head installed, M3 will simply turn it on. The "S" 18000 will simply be ignored.) Line number

3 is simply a command to "rapid" quickly to the home location of $X = 0$ $Y = 0$. Line N4 uses the M98 command to call (i.e., P1000) the subroutine labeled O1000. Hence, now the focus of processing is in the subroutine and stays there until it reaches the M99 command, indicating the end of the subroutine. Processing focus then returns to the main, but at the next line (N5). Then, line N5 disengages the spindle head and M30 on line N6 instructs the program to rewind to the beginning of the main file and end or halt all program operation.

Example 2

We now delve a bit deeper with the same file used as in example 1. However, we will now use the *main* file that calls the same *subroutine* for execution in four different locations on the CNC work table.

```
**Main Program**
N1      G90 G20 G49     (this is an initialization string)
N2      M03 S18000      (turn on the spindle head)
N3      G0 X0 Y0        (rapid move to the G54 home location of 0,0)
N4      G52 X0 Y0       (first temporary part location that also
                            corresponds to G54 0,0 location)
N5      M98 P1000       (call subroutine number 1000)
N6      G52 X5 Y0       (second temporary part location at new location)
N7      M98 P1000       (call subroutine number 1000)
N8      G52 X5 Y5       (third temporary part location at new location)
N9      M98 P1000       (call subroutine number 1000)
N10     G52 X0 Y5       (fourth temporary part location at new location)
N11     M98 P1000       (call subroutine number 1000)
N12     G52 X0 Y0       (cancel using temporary locations)
N13     M05             (turn off the spindle head)
N14     M30             (program rewind and end)

**Subroutine**
O1000    (subroutine label 1000. Note the "O" is a letter, not number zero.)
*Start* of cutting/engraving blocks of G code
   ...           (this area is where the programmer would add either manually)
   ...           (written or from CAM the lines of G code that get used in each)
   ...           (instance at each temporary part location)
   ...
   ...
   ...
*End* of cutting/engraving blocks of G code
M99      (exit subroutine and return to next line of code in the main)
```

Let's take a closer look at example 2 to discuss what is happening. Again, example 2 is similar to the first example; however, it has some G52 commands inserted thus allowing *temporary offset* locations. Temporary offset locations allow the same execution of code to be performed at specified locations within the working envelope.

Line number 3 is simply a command to "rapid" quickly to the location that is set at G54 X0 and Y0 for preparation of getting into a "temporary offset" mode (i.e., G52). Then, on N4, we are now at the starting temporary offset location of X0 and Y0. This is the first "temporary part location" to be worked at. N5 then calls a subroutine with the label O1000. Note P is the subroutine calling identifier and the letter O denotes the subroutines label. It is the numeric value associated with the P and O that allow multiple subroutines to be called (such as, O1001, O1002, etc.). Once the M98 P1000 command is encountered on line N5, the program focus leaves the main routine and goes to the O1000 subroutine and continues to process whatever code is contained within it. Once the M99 command is reached in the subroutine, the focus exits the subroutine and returns focus to the next line in the main routine. In this case, N6 is the next line to be executed. Next, line N6 instructs the program to use another temporary offset at location $X = 5$ and $Y = 0$. Then, line N7 (just as was in the case of line N5) calls the same O1000 subroutine.

In short, whatever is contained in subroutine O1000 will be performed at X0 Y0, then at X5 Y0, then X5 Y5, and, finally, X0 Y5 making a counterclockwise loop starting from the lower left corner. When all of the locations have been addressed, line N12 commands a G52 at location X0 Y0 to clear the logic of any temporary work offsets (i.e. back in G54 mode). Line N13 turns off the spindle head and the final line containing M30 issues a rewind and end the main program that was executing. Also of importance is to note that subroutines can either be part of the same file that the main is in or can be a different file altogether. In these examples, the subroutines are shown as being included in the same file as the main routine. Then, too, it is possible to call another subroutine from within an existing subroutine. Typical nesting of this type generally extends to four levels of logic. You can, however, have an unlimited number of subroutines as part or separate to the main program. As an exercise left to the reader, two or more of the same files can be easily modified for cases where multiple tooling is required. In such an instance, each of the main routines would be identical with the exception of the "P" calling labels for the different subroutines. For example, main file 1 would use M98 P1000 to call subroutine O1000. Main file 2 would use M98 P2000 to call its corresponding O2000 subroutine. This same approach can be used as many times as tooling changes necessitates.

Using subroutines, as applicable, can greatly minimize your file sizes and is a great introduction to learning G-code programming. There are a great many more types of features that G coding has to offer (such as several canned routines) that are beyond the scope of this text to cover. I do, however, sincerely hope the reader will not be intimidated by this "seemingly cryptic" language and explore the many benefits it has to offer the CNC user.

APPENDIX C
Engineering Process of Selecting a Ball Screw

In the transmission section of this book, the process of selecting the type of rack and pinion system was covered. Also in the same section were discussions regarding both lead and ball screws. In this addendum, a detailed process as to the selection of the best ball screw for you to choose and use is presented. Please note that although this selection process is specific to those of ball screws, there is also a similar engineering guide available through Thomson Linear for lead screws and their regular and anti-backlash nuts (see www.thomsonlinear.com). Special credit for the following ball screw product selection overview goes to Thomson Industries, Inc.

Appendix C

Precision Rolled Ball Screws Product Overview — Inch Series

Tangential Ball Return
A unique Thomson feature which minimizes recirculated bearing ball deflection, for smoother and quieter operation. The tangential circuit consists of a pick-up deflector finger and modified return tube which allows the bearing balls to enter and exit the load-carrying portion of the ball screw circuit in a straight path. Standard on ball bearing screws with up to 10,000 pound dynamic load capabilities.

Load Locking Spring
The load locking spring is a coil turned into the inactive portion of the nut and conforms to the ball tract. In normal operation, the spring is inactive and not in contact with the screw. In the event the ball bearings are lost from the nut, the load locking spring will not allow the load carrying nut to free-fall down the screw.

End Journals and Bearing Supports
To assist the designer, standard end journals and bearing supports are included in this catalog. Ball screw assemblies, complete with end journals and bearing supports, may be ordered through a local Thomson distributor or directly from the factory.

Thomson welcomes the opportunity to custom machine end journals to unique customer designs.

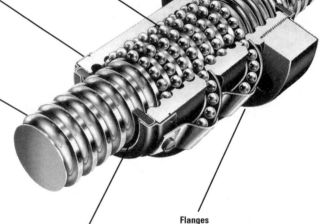

Lube Holes
A standard 1/8-27 NPT tapped hole on ball nuts with a dynamic load capacity of 10,000 pounds or more provides easy access for continuing lubrication.

Coating
Our catalog standard ball screws feature a high luster polished and oiled finish, which provides superior surface finish, smoother operation, and a high quality look and feel which is consistent across our entire product line. Additional ball screw coatings (thin dense chrome, black oxide, manganese phosphate) are available upon request.

Wiper Kit
Wipers can increase the life and long-term performance of ball bearing screws by preventing most dirt and other foreign matter from entering the ball nut. Wipers are attached via two methods: Type A attaches directly to the ball nut body and flange; and Type B installs into the ends of the ball nut with easy-to-install snap rings kits. See our installation section on page 227 for more details.

Flanges
Standard flanges are offered for all ball nuts. Flanges provide an easy, low-cost method to mount the load square and concentric to the ball screw.

Type A

Type B

Engineering Process of Selecting a Ball Screw

Selecting a Ball Screw Assembly for Your Application — Inch Series

A ball screw assembly is a mechanical device for translating rotational motion to linear motion. As well as being able to apply or withstand high thrust loads, they can do so with minimum internal friction. They are made to close tolerances and are therefore suitable for use in situations in which high precision is necessary. The selection of the correct ball screw assembly for a specific application is an iterative process to determine the smallest envelope and most cost-effective solution. Below is a list of the most common (but not complete) design considerations used to select a ball screw assembly.

- Compression or Tension Load
- Linear Velocity
- Positional Accuracy and Repeatability
- Required Life Expectancy
- Mounting Configuration
- Dimensional Constraints
- Input Power Requirements
- Environmental Condition

At a minimum, the design load, linear velocity, and positional accuracy should be the known inputs and are used to calculate the diameter, lead, and load capacity of the ball screw assembly. Individual ball screw components are then selected based on life, dimensional constraints, mounting configuration, and environmental conditions.

The following procedure will take you through the most common application-based selection of a ball screw assembly. As no two applications are the same, so the determination process is never the same.

1. Determine the required positional accuracy and repeatability that your application requires (page 202). Backlash is the linear independent motion between the ball screw and the ball nut and can be controlled by preloading the ball nut (page 203). The manufacturing process, rolled screws versus ground screws, dictates the accuracy (page 203).
2. Determine how you plan to mount the ball screw assembly into your machine (see page 205). The configuration of the end supports and the travel distance (Max L) will dictate the load and speed limitations of the ball screw.
3. A ball nut in tension can handle loads up to the rated capacity of the nut. For a ball nut in compression, calculate the Permissible Compression Loading (page 201) or use the Compression Loading Chart (page 209) to select a ball screw diameter that meets or exceeds your design load.
4. Calculate the lead of the ball screw that will produce the speed requirement (page 200).
5. The ball nut life can then be calculated using the Dynamic Load Rating (C_{am}) provided in the catalog detail pages (page 200) or use the Life Expectancy Charts (pages 207 or 208).
6. Every ball screw has a rotation speed limit, which is the point of excessive vibration/harmonics in the screw. The critical speed is dependent on the end support configuration. Calculate the Critical Screw Speed of the chosen ball screw (page 201) or use the Acceptable Speed Chart (page 206) to determine the critical speed.
7. If the load, life and speed calculations confirm that the selected ball screw assembly meets or exceeds the design requirements, then proceed to the next step. If not... Larger diameter screws will increase the load capacity and increase the speed rating. Smaller lead screws will decrease the linear speed (assuming constant input motor speed), increase the motor speed (assuming constant linear speed), and decrease the input torque required. Higher lead screws will increase the linear speed (assuming constant input motor speed), decrease the input motor speed (assuming constant linear speed), and increase the input torque required. Repeat steps 3 through 5 until the correct solution is obtained.
8. Determine how the ball nut will interface into your application. A ball nut flange is the typical method of attaching the ball nut to the load. Threaded ball nuts and cylindrical ball nuts are alternative ways to provide the interface.
9. Additional design considerations and features are also available. Preloaded ball nuts are available to reduce system backlash and increase positional accuracy. Wiper kits to protect the assembly from contaminants and to contain lubrication are standard on some units and optional on most others. Bearing supports and end machining are also available as options for all ball screws.
10. The final considerations are system mounting and lubrication. The ball nut should be loaded axially only as any radial loading significantly reduces the performance of the assembly (page 204). The assembly should also be properly aligned with the drive system, bearing supports, and load to achieve optimal performance (page 204). The ball screw assembly should never be run without proper lubrication. Many lubricants are available depending on the application and environment (page 204).

Note: Application and customer service support is available to assist in the selection of your ball screw assembly. Please contact your local Danaher representative or the customer support center (1-540-633-3549 — DMAC) for any additional assistance.

Appendix C

Ball Screw Assembly Selection Example:
Inputs:
  Load: 30,000 lb. Compression Maximum
      10,000 lb. dynamic
  Linear Speed: 200 in./min.
  Input Speed: 400 rpm
  Travel: 85 in.
  Life: 2 x 10⁶ inches

1. **Accuracy (pages 202 and 203)**
   No Preload and Standard Rolled (±.004 in./12 in.)

2. **End Supports (page 205)**
   Fixed/Supported

3. **Determine Screw Diameter**
   From Chart (page 209): Ø2.000 in.

   From Equation (page 201): $30,000 / .8 = \dfrac{1.47 \times 1.405 \times 10^7 \times d_r^4}{(85)^2}$

   therefore, $d_r = 1.903$ in.

4. **Determine Lead (page 200)**

   Lead $= \dfrac{200 \text{ in./min.}}{400 \text{ rpm}}$ therefore, Lead = .500 in.

5. **Determine Life**
   From Catalog (page 77): Dynamic Load = 18,500 lbs.

   From Equation (page 200): Life (inches) = $\left[\dfrac{18,500}{10,000}\right]^3 \times 10^6$

   therefore, Life = 6.3×10^6 inches

   Verified via Chart (page 207)

6. **Determine Critical Speed**
   From Catalog (page 77): Screw Root Diameter is 1.72 in.

   From Equation (page 201): $.8 \times 1.47 \times 4.76 \times 10^6 \times \dfrac{d_r}{l^2}$

   therefore, Speed = 1,332.6 rpm

   Verified via Chart (page 206)

7. **Design Verification**
   OK per load, speed and life.

8. **Load Interface**
   Flanged connection preferred.

9. **Additional Requirements**
   - Wipers required
   - Bearing Supports required
   - End Machining needed
   - Right Hand Thread
   - Carbon Steel

10. **Mounting and Lubrication**
    System will require motor interface and linear rails for alignment.
    TriGel 450R

Product Selection (page 77):
  Ball Nut: P/N 8120-448-011
  Ball Screw: P/N 190-9112
  Wiper Kit: P/N 8120-101-002
  Flange: P/N 8120-448-002

Engineering Process of Selecting a Ball Screw

Design Formulas

These formulas allow you to calculate a number of important factors which govern the application of Thomson ball screws.

1. Ball Screw Life (L)

The ball screw assembly's useful life will vary according to load and speed. Life is typically rated at 90% confidence, L10 (which represents time at which 90% of assemblies still perform).

Functional life should be determined by approximating equivalent rotational speed and loading force over typical performance cycles.

Simple rotational speed profile

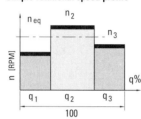

Simple loading profile (1)

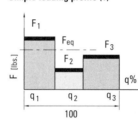

Simple loading profile (2)

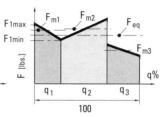

$$n_{eq}[\min^{-1}] = \sum_{i=1}^{n} n_i \times \frac{q_i}{100}$$

$$F_{eq}[\text{lbs.}] = \left(\sum_{i=1}^{n} F_i^3 \times \frac{n_i}{n_{eq}} \times \frac{q_i}{100} \right)^{1/3}$$

$$F_{eq}[\text{lbs.}] = \left(\sum_{i=1}^{n} F_{mi}^3 \times \frac{n_i}{n_{eq}} \times \frac{q_i}{100} \right)^{1/3}$$

Modified Life

$$L_{10}[\text{inches}] = \left[\frac{C_{am}}{F_{eq}} \right]^3 \times 10^6$$

$$L_{h10}[\text{hours}] = \frac{L_{10}}{n_{eq} \times 60}$$

Parameters:

- n_{eq} = Travel Rate (inches/min)
- F_{eq} = equivalent operating load [lbs.]
- C_{am} = dynamic load rating [lbs.] (see product detail pages) (Based on 1.0 million inches)

2. Rotational Speed Required for a Specific Linear Velocity

$$n = \frac{\text{Travel Rate (in. x min.}^{-1})}{\text{Lead (in.)}} \qquad n = \text{rpm}$$

3. Machine Service Life

After ball screw life (L) is calculated, apply it to the following formula to determine machine service life.

$$\text{Machine Service Life (in years)} = \frac{L_{h10}[\text{hours}]}{(\text{machine operating hours}) \cdot (\text{days/year}) \cdot \left(\frac{\text{ball screw operating hours}}{\text{machine operating hours}} \right)}$$

Appendix C

4. **Torque**

   a. Driving torque: T_d (lb$_f$-in.) = $\dfrac{F_{eq} \times P}{2\pi e}$ = $0.177 \times F_{eq} \times P$

   b. Backdrive torque: T_b (lb$_f$-in.) = $\dfrac{F_{eq} \times P \times e}{2\pi}$ = $0.143 \times F_{eq} \times P$

   F_{eq} = Equivalent Operating Load (lb$_f$)
   P = Lead (in.)
   e = Efficiency = 0.90
   T_d = Driving Torque (lb$_f$-in.)
   T_b = Backdrive Torque (lb$_f$-in.)
   1 lb$_f$-in. = 0.113 (N•m)

   (conversion of linear to rotational motion)

5. **Power**

   P_d (hp) = $\dfrac{F_{eq} \times P}{(2\pi) e} \times \dfrac{n}{6.3021 \times 10^4}$ = $\dfrac{F_{eq} \times P \times n}{3.564 \times 10^5}$

   P_d = Power (hp)
   n = rpm
   1 hp = 746 W

6. **Permissible Rotational Speed**

   The permissible rotational speed depends on two factors: critical screw speed and critical nut speed.

   6a. Critical Screw Speed

   The critical screw speed is related to the natural frequency of the screw shaft. Exceeding this value may result in excessive vibration. The critical screw speed may be found using the following equations or the chart on page 206.

   $n_c = C_s \times 4.76 \times 10^6 \times \dfrac{d_r}{l^2}$

   $n_s = n_c \times S$

   n_c = Critical Speed (rpm)
   n_s = Safe Drive Speed
   d_r = Root Diameter (in.)
   l = Length between Bearing Supports (in.)
   S = Safety Factor (0.8 maximum)
   C_s = End Fixity Factor

End Fixity Factor - Critical Screw Speed		
End Supports		C_s
A	One end fixed, one end free	0.36
B	Both ends supported	1.00
C	One end fixed, one end supported	1.47
D	Both ends fixed	2.23

   6b. Critical Nut Speed

   The critical nut speed is related to the velocity of the ball bearings rotating around the screw shaft. Exceeding this value may result in permanent damage to the ball recirculation components. Thomson recommends a maximum DN value of 3000 for standard tube transfer designs with a lead-to-diameter ratio less than 2/3. For sizes with a lead-to-diameter ratio of 2/3 or greater such as .750 x .500 and 1.500 x 2.000, the product detail pages should be consulted to find the recommended maximum speed.

   $DN = d_0 n$

   where
   d_0 = nominal shaft diameter (in)
   n = rotational speed of shaft (rpm)

7. **Permissible Compression Loading**

   Exceeding the recommended maximum compression force may result in buckling of the screw shaft.

   $F_c = \dfrac{C_s \times 1.405 \times 10^7 \times d_r^4}{l^2}$

   $F_s = F_c \times S$

   F_c = Critical Buckling Force (lbs.)
   F_s = Safe Compression Force (lbs.)
   d_r = Root Diameter (in.)
   l = Max Unsupported Length (in.)
   S = Safety Factor (0.8 maximum)
   C_s = End Fixity Factor

End Fixity Factor - Permissible Compression Loading		
End Supports		C_s
A	One end fixed, one end free	0.25
B	Both ends supported	1.00
C	One end fixed, one end supported	2.00
D	Both ends fixed	4.00

Accuracy Classes

Accuracy is a measure of how closely a motion system will approach a command position. Perfect accuracy, for example, means that advancing a ball nut a precise amount from a given point on the screw always requires exactly the theoretically predicted number of revolutions.

Inch ball screws are produced in two main tolerance classes: Precision and Precision Plus. Precision grade ball screws are used in applications requiring only coarse movement or those utilizing linear feedback for position location. As such, most Precision grade screws are provided with nuts having backlash. Precision Plus grade ball screws are used where repeatable positioning within microns is critical, without the use of a linear feedback device.

Differences between Precision and Precision Plus grades are highlighted in the graph. Precision grade screws allow greater cumulative variation over the useful length of the screw. Precision Plus grade screws contain accumulation of lead error to provide precise positioning over the screw's entire useful length.

Precision Plus Ball Screws

Maximum error over useful length = $e_p + 1/2V_{up} + C$

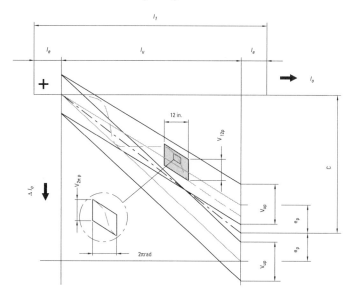

l_o = nominal travel
l_t = thread length
l_o = travel deviation
l_u = useful travel
l_e = excess travel
C = travel compensation for useful travel (std. = 0)
e_p = tolerance for actual mean travel deviation (the difference between the maximum and minimum values of the permissible actual mean travel)
V_{up} = permissible travel variation within useful travel, l_u
V_{12p} = permissible travel deviation within 12-inch travel
$V_{2?p}$ = permissible travel deviation within 1 revolution

Precision Ball Screws

Maximum error over useful length = e_p

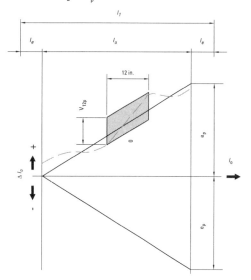

Appendix C

Permissible Travel Variation Over Usable Length

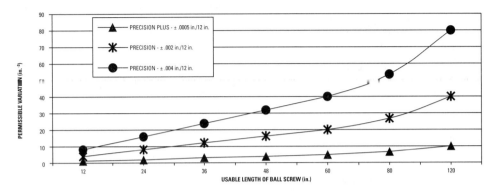

Tolerance Class	Lead Accuracy V_{300p}		Permissible Travel Deviation V_{up} (in.$^{-3}$) Over Screw Length l_u (in.)						
		$l_u =$	12	24	36	48	60	80	120
Precision Plus*	±.0005 in./12 in.	V_{up} (in.)	1	2	3	4	5	6.67	10
Precision	±.002 in./12 in.	V_{up} (in.)	4	8	12	16	20	26.7	40
Precision*	±.004 in./12 in.	V_{up} (in.)	8	16	24	32	40	53.3	80

* Standard product tolerances

Preload Types

Skip-Lead Preload

- The lead is offset within the ball nut to provide a precise preload.
- Typically used where both repeatability and high stiffness are required.

Double-Nut Adjustable Preload

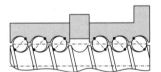

- A compression spring is used to axially load two ball nuts against each other.
- Typically used for positioning applications where repeatability is critical.

No Preload

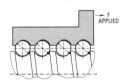

- Axial play is present between screw and nut (typically .002"–.008" depending on size).
- Typically used for transport or vertical applications.

Engineering Process of Selecting a Ball Screw

Lubrication Guidelines

Ball screws must be lubricated to operate properly and achieve the rated life. We recommend using TriGEL-450R or TriGEL-1800RC for lubricating ball screws. Other oils and greases may be applicable but have not been evaluated.

The TriGEL grease can be applied directly to the screw threads near the root of the ball track. Some ball nut sizes are available with threaded lube holes for mounting lubrication fittings. For these ball nuts, the TriGEL grease can be pumped directly into the nut. Please refer to the catalog detail views to verify which ball nuts have the threaded lube holes. It is recommended to use these nuts in conjunction with a wiper kit to contain the lubricant in the body of the nut.

Ball screws may require lubrication frequently, depending on both environmental and operating conditions. If the lubricant appears to be dispersed before this point or has become dry or crusted, the maintenance interval should be reduced. Before adding additional grease, wipe the screw clean, removing the old grease and any particulate contamination seen on the screw. If oil is being used, the best results may be obtained by utilizing a continuous-drip type applicator.

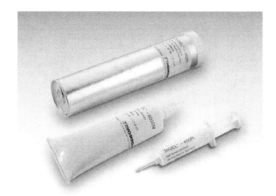

Nut Loading

Axial loading (on nut or screw) is optimal for performance and life.
For applications requiring radial loads, please contact us.

Axial Loading: optimal

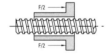 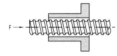

Radial Loading: detrimental*

  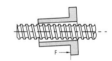

* Minimize radial loading to less than 5% of the axial load.

Nut Mounting

Use the following guidelines to achieve optimal performance.

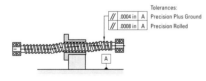

Tolerances:
| // | .0004 in | A | Precision Plus Ground |
| // | .0008 in | A | Precision Rolled |

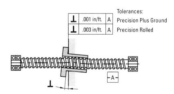

Tolerances:
| ⊥ | .001 in/ft. | A | Precision Plus Ground |
| ⊥ | .003 in/ft. | A | Precision Rolled |

Appendix C

Bearing Support Reference Drawings (End Fixity)

Critical Speed — That condition where the rotary speed of the assembly sets up harmonic vibrations. (Refer to Figure 1.) These vibrations are the result of shaft diameter, unsupported length, type of bearing support, position of the ball nut in the stroke, how the ball nut is mounted, the shaft or ball nut rpm, etc. (Note: Shaft vibrations may also be caused by a bent screw or faulty installation alignment.) The four end fixity drawings (A, B, C, and D) show the bearing configurations for supporting a rotating shaft. The selection chart for Travel Rate vs. Length on page 206, shows these same configurations at the bottom of the chart and factors in their effect on critical shaft speed for the unsupported screw length.

Figure 1

Bearing Support vs. Speed (travel rate or rpm)
A – One end fixed, other end free
B – Both ends supported
C – One end fixed, other end supported
D – Both ends fixed

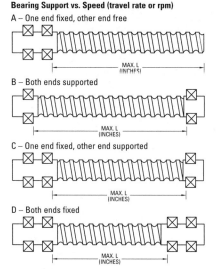

Tension Loads — Those loads where the force pulls on the bearing and its support. (Refer to Figure 2.) Where practical, applications should be designed to function with the load in tension to achieve the widest possible selection of screw sizes. Ball screws operating in both tension and compression may be preloaded between the support bearings or mounted per the guidelines under Compression Loads.

Figure 2

Compression Loads — Those loads where the force pushes on the bearing and its support. (Refer to Figure 3.) Compression loads tend to cause the screw shaft to bend. This normally requires a ball screw with a larger diameter than one for tension loading only. The four end fixity drawings (A, B, C and D) show the bearing configurations for supporting a shaft subject to compression loads. The selection chart for Compression Load vs. Length, on page 209, shows these same configurations at the bottom of the chart and factors in their effect on the unsupported length of the screw for compression loads.

Figure 3

Bearing Support vs. Compression Load on Screws
A – One end fixed, other end free
B – Both ends supported
C – One end fixed, other end supported
D – Both ends fixed

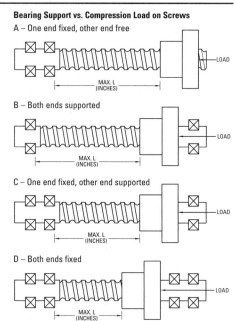

Note: The information in this guide for end fixity is based on the centers of the two bearings spaced apart by 1-1/2 times the root diameter of the screw.

Engineering Process of Selecting a Ball Screw

Acceptable Speed† vs. Length for Screws

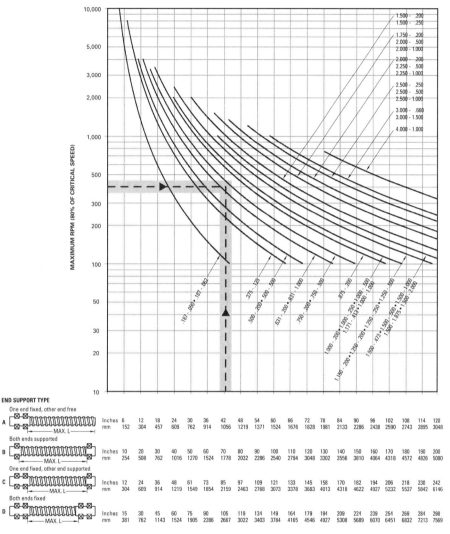

Example: Travel rate of 400 rpm.
Unsupported length of 85 in. (2159 mm).
End fixity of one end fixed, other end supported.

All screws with curves which pass through or above and to the right of the plotted point are suitable for the example. The acceptable velocities shown by this graph apply to the screw shaft selected and are not indicative of the velocities attainable of all of the associated ball nut assemblies. Consult Thomson Engineering for high speed applications.

†80% of critical speed

Appendix C

Life Expectancy for Precision Ball Screw Assemblies

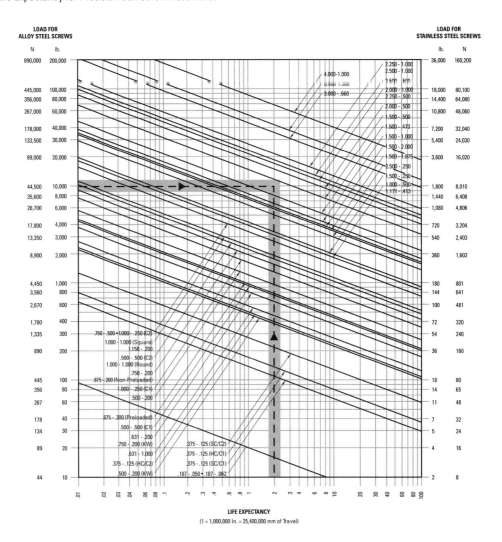

C1 = Single Circuit   C2 = Double Circuit   SC = Standard Capacity   HC = High Capacity

Example: Application life expectancy (total travel) desired is 2 million in. (50.8 million mm). Normal operating load is 10,000 lb. (44,500 N).

All screws with curves which pass through or are above and to the right of the plotted point are suitable for the example. The suitable dynamic life expectancies shown in this graph are not to exceed the maximum static load capacity as given in the rating table for the individual ball nut assembly.

Engineering Process of Selecting a Ball Screw   245

Life Expectancy for Precision Plus Preloaded Ball Screw Assemblies

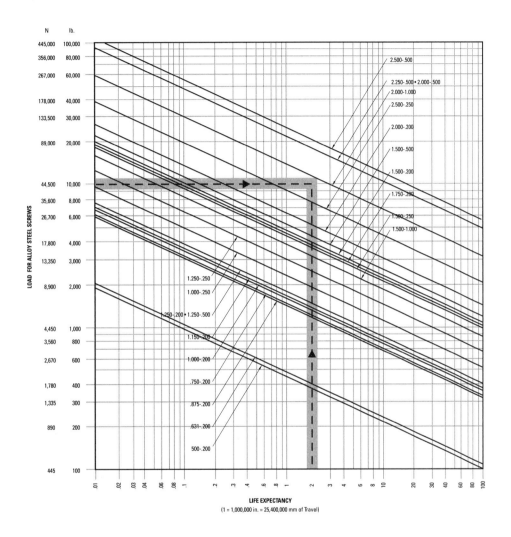

Example: Application life expectancy (total travel) desired is 2 million in. (50.8 million mm).
Normal operating load is 10,000 lb. (44,500 N).

All screws with curves which pass through or are above and to the right of the plotted point are suitable for the example. The suitable dynamic life expectancies shown in this graph are not to exceed the maximum static load capacity as given in the rating table for the individual ball nut assembly.

Appendix C

Compression Load vs. Length for Designated Ball Screws

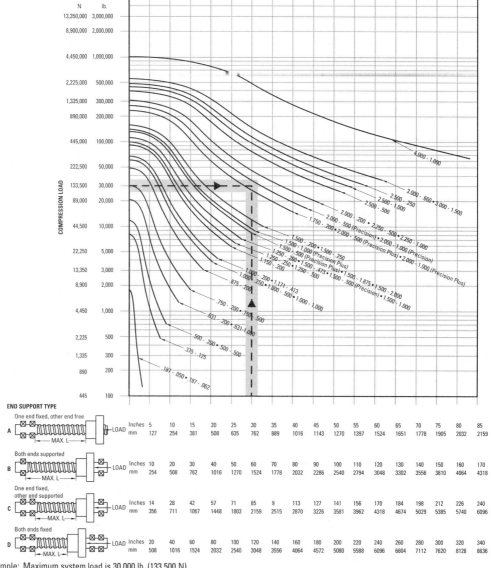

Example: Maximum system load is 30,000 lb. (133,500 N).
Length of 85 in. (2159 mm).
End fixity of one end fixed, other end supported.

All screws with curves which pass through or above and to the right of the plotted point are suitable for the example.

The suitable compression loads shown in this graph are not to exceed the maximum static load capacity as given in the rating table for the individual ball nut assembly.

APPENDIX D
NEMA Motor Mounting Templates

If you are interested in building your own CNC machine or fabricating some reduction units for an existing machine, you will need a mounting pattern for the size motor you are planning to use. For whatever reason, NEMA foot print patterns are not something that is readily available on the Internet or in most books. Hence, included here are three drawings of some commonly used motor sizes. I will mention that there can and will be minor differences in dimensions from manufacturer to manufacturer, but those supplied here were taken by the author using some motors I had in my shop that parallel the same dimensions from most other motors I have used over the years. The placement of the mounting holes and the overall frame size of the motor are deemed universal so you can use them in your designs.

Typical discrepancies in dimensions between those listed here and those available from various manufacturers can include: output shaft diameter, the diameter of the raised-boss circular area located on the base of the mount, and the actual diameter of the four mounting holes. If the mounting holes on your motors are too small (often a metric size), it is perfectly acceptable to drill or ream the hole to the size required. If the diameter of the motor's output shaft is metric, you will need to purchased metric-sized pulleys or shaft couplers as changing the diameter is not an easy or inexpensive task.

FIGURE D-1
NEMA 17 footprint.

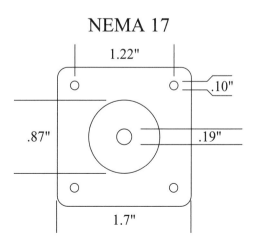

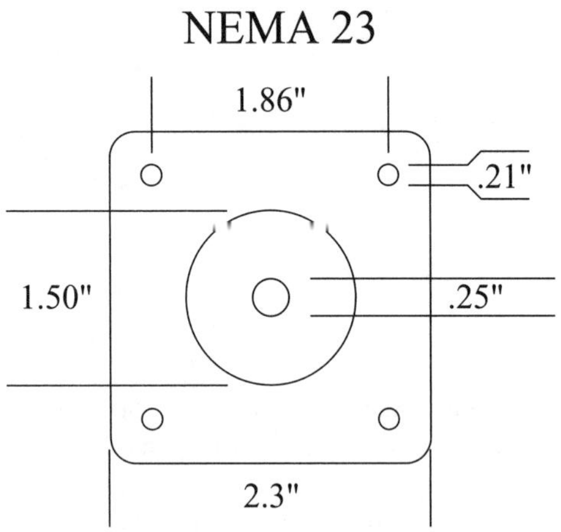

Figure D-2
NEMA 23 footprint.

Figure D-3
NEMA 34 footprint.

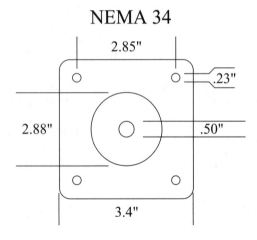

Index

Note: Page numbers followed by f and t indicate figures and tables.

A

Absolute distance mode (G90), 112–113, 113f
Absolute IJ mode, 97
Accented fonts, 197–198, 198f
Acme screws, 38
 teeth per inch (TPI), 41
ADA (American Disabilities Act), 21
ADA signage, 192–194
 ADA restroom sign, 193f
 ADA sign, 194f
Adjunct devices for controller hardware, 78–79
Adobe Illustrator, 140
Aftermarket engraving spindle, 10, 11f
Alpha-numeric ASCII-based machine-command, 90–93, 92t–93t
Alternating current (AC) motors, 62
Aluminum, 12, 18–19
American Disabilities Act, 21. *See also* ADA signage
Angle bracket (with lead screw nut), 180f
Angular contact bearings, 35
Antares, 21
Antibacklash nut, 38, 39f, 40f
Arc at feed rate (G02, G03), 96–99, 96f
ArtCAM (by DelCAM), 155
Assembly with carriage traveler plate, 180f
ATC. *See* Automatic tool changer
AutoCAD, 138–139
AutoCAD-LT (light), 138–139
AutoEditNC, 90
Automatic tool changer, 5, 10

B

B and B Plasma (company), 166, 166f
Background texture, 197, 197f
Backlash, 38
Backlit acrylic LED sign, 201, 202f
Ball nose, 20
Ball screw(s), 41–42
 selection. *See* Precision rolled ball screws product overview
Ball screw life, 237
Bearing removal tool, 8
Belt reduction unit, 51f
Belt tensioning tool, 52f
Bevel-edge lettering, 195–196, 196f
BHCS. *See* Button head cap screws
Bipolar series wiring, 58
BOB. *See* Breakout board
Braille lettering, 21, 193–194
Breakout board, 72–75. *See also under* Controller hardware
Button head cap screws, 178

C

CAD and graphics, 135
 graphics programs

251

CAD and graphics *(continued)*
   Adobe Illustrator, 140
   GIMP, 140
   Rhino, 140
   SolidWorks, 140
  pixilation, 137f
  raster to vector conversion utilities, 137–138
  stick font text, 136, 136f
  TrueType text, 136, 136f
  2D and 3D, difference between, 138
  vendors, listing of, 138
   AutoCAD, 138–139
   AutoCAD-LT, 138–139
   free CAD, 139
   low-cost CAD, 139
   QCAD, 139–140
   TurboCAD, 139
CAM software, 143, 156
  CAD and CAM combination software
   ArtCAM (by DelCAM), 155
   CAM-Only Software, 156
   FreeMill, 156
   MasterCAM, 156
   SheetCAM (by Stable Design), 156
   Vectric Software, 155
   VisualMill, 156
  milling options, generalized
   climb milling, 152–153
   conventional milling, 152–153
   finish profiling, 153–154
   lead in position, 154
   lead out position, 154
   nesting, 154–155
   part tabbing, 154
   ramped entry cut, 154
   step-down, 153
   step-over, 153
  uses of, 144–146
   area clearance, 147
   CAM machining parameters, 146
   cuts, decorative types of, 148
   drilling function, 148–149
   incised carving, 150–151
   inlay function, 149
   profiling, 146–147, 145f, 146f
   3D carving, 150–151
Campaign buttons, 201

Carriage(s), 28
  assembly, 31, 32f
  with $4^3/_4$-in riser tubing, 177f
Cartesian coordinate system
  definition, 129–130, 129f
  lathe/rotary topology, 131–132, 131f, 132f
  table or mill topology, 130–131, 131f
C-channel, 19
Ceramic-style bearings, 8
Clamp-on mechanism, 37
Clam-shell method, 184
Climb milling, 152–153
Closed bearing style, 27
CNC
  engraving machine, 193
  plasma table, 15
   offered by B and B Plasma, 166f
CNC machines
  construction materials (router/plasma)
   aluminum, 18–19
   joining materials together, 19
   wood composites, 18
  engraver, 5
  hold-down methods
   double-sided tape, 12
   dual Z, 15–16
   fat mat, 12
   limit and homing switches, 16
   mechanical probing, 13
   optical probing, 13
   plasma cutters, 14–15
   probing, 13
   rotary *A* axis, 13–14
   T track grid work, 12
   vacuum, 12
  lathes, 18
  mills
   accuracy versus repeatability, 17–18
  resolution, 11
  router, 5
  router head, 6–8
  spindle head, 8–11
  tooling, 19–21, 21–22
CNC plasma table, 173, 175f
  complete leg bottom, 174f

Index

completed table, 183f, 184f
gantry end caps, 177, 177f
leg members, 175f
leg replacement, 175f
leg-leveling system, 173, 174f
materials assembled for, 174f
table assembly, 184
X-axis members, 175f
Z-axis assembly, 181f, 182f
CNC router, items produced on
accented fonts, 197–198, 198f
ADA signage, 192–194
ADA restroom sign, 193f
ADA sign, 194f
backlit acrylic LED sign, 201, 202f
bevel-edge lettering, 195–196, 196f
engraving using engraving stock, 205, 206f
filled-shallow engraving, 197
furniture, 189
shop cabinetry, 189f
monument sign, 189, 190f
musical instrument. *See* Solid-body electric guitar
oversized projects, 200, 201f
panel with removed lettering, 204–205, 205f
programming examples, 219–223
pulley blocks, 198–199, 199f, 200f
raised/recessed-lettering effects, 196, 196f
relay cover removal job, 216–218, 218f, 220f
stud-mounted lettering and logos, 214–216, 215f, 216f
textured background effect, 197, 197f
3D routing, 202–204, 203f
tools and templates, 189
dovetail jig, 190f, 191f
DT joint cut, 192f
fret wire bending tool, 193f
V carving, 194–195, 195f
Coating, ball screw, 234
Compression loads, 242
Computer operating systems and applications, 89
Concentric bushing, 30, 30f
Conical tooling, 20f, 21f
Construction materials. *See under* CNC machines
Control software
computer operating systems and applications, 89
enhanced machine controller, version 2 (EMC2), 88–89
feed rate (F), 123
G code, 90–93, 92t–93t
G00 (rapid linear motion), 93–95
G01 (linear motion at feed rate), 95–96
G02 and G03 (arc at feed rate), 96–99
G04 (dwell), 99
G10 (coordinate system data tool and work offset tables), 99–100
G12 and G13 (clockwise/counterclockwise circular pocket), 100
G15 and G16 (exit and enter polar mode), 100
G17, G18, and G19 (plane selection), 100, 100f, 101f
G20 and G21 (length units), 101
G28 and G30 (return to home), 101
G28.1 (reference axes), 101–102
G31 (straight probe), 102–103
G40, G41, and G42 (cutter radius compensation), 103–104
G43, G44, and G49 (tool length offsets), 104
G50 and G51 (scale factors), 104
G52 (temporary coordinate system offset), 105
G53 (move in absolute coordinates), 105
G54 to G59 and G59 P (select work offset coordinate system), 105
G61 and G64 (path control mode), 106
G68 and G69 (coordinate system rotation), 106
G73 canned cycle (High-Speed Peck Drill), 106–107
G80 (modal motion cancellation), 107
G81 cycle, 109
G81 to G89 (canned cycles), 108–109, 110f
G82 cycle (for drilling), 110

Control software *(continued)*
  G83 cycle (peck drilling), 110
  G85 cycle (for boring or reaming), 111, 111f
  G86 cycle (for boring), 112
  G88 cycle (for boring), 112
  G89 cycle (for boring), 112
  G90 and G91 (distance mode), 112–113
  G92, G92.1, G92.2, and G92.3 – G92 Offsets, 113–114
  G93, G94, and G95 (path control mode set), 114–115
  G98 and G99 (canned cycle return level), 115–116
  G-Code editors, 90
  M codes, 116, 116t
  M0, M1, M2, and M30 (program stopping and ending), 116–117
  M3, M4, and M5 (spindle control), 117
  M47 (rerun from first line), 118
  M48 and M49 (override control), 118
  M6 (tool change), 117–118
  M7, M8, and M9 (coolant control), 118
  M98 (subroutine call), 118–119
  M99 (return from subroutine), 119
  M998 (tool change position), 121
  Mach3 Control Software, 87–88
  modal groups, 122, 122t
  select tool (T), 123
  self-reversing tapping cycles, 119–121, 120t–121t
  spindle speed (S), 123
  user-defined M codes, 122
Controller hardware
  adjunct devices, 78–79
    spindle-speed controller, 79
  breakout board, 72–75
    dual-port BOB with spindle-speed controller, 79f
    dual port from sound logic, 75f
    five axis multipurpose, 74f
    four-axis BOB from PMDX, 74f
    functions of, 72
  drives, 75–76
  enclosure, 71–72, 72f
    wiring, 73f
  pendant, 79–81
    micro switches, 81–83
    Z touch-off tool, 83
  power supply, 76–78, 77f, 78f
    wiring, 83
Controller software, 163
  console, 10
Conventional milling, 152–153
Conversational programming, 88
Coolant control, 118
CorelTrace, 137
Critical nut speed, 238
Critical speed, 242
Current spikes, 7
Cut relay in tray socket, 219f
Cuts, decorative types of, 148
Cutter radius compensation (G40, G41, and G42), 103–104, 105
Cutter ramping, 154
Cutter step-downs, 153

D

Digital readout device, 65
DRO device. *See* Digital readout device
Direct current motors, 62
DC motors. *See* Direct current motors
Do-it-yourself (DIY), 168
Double-nut adjustable preload, 240
Double-sided masking tape, 12
Dovetail jig, 190f, 191f
Drilling function, 148–149
Drives, 75–76
Driving method, 37
Dual Z, 15–16
Duality lathe, 169

E

Eccentric bushing, 30, 30f
EMC2. *See* Enhanced machine controller, Version 2
Emergency power off (EPO), 71, 72
Encoders, 64-65, 65f
  assembly, 65f

with dual-shaft configuration, 64f
  linear glass scales, 65
Enclosure, 71–72, 72f
End fixity, 35–36
End journals and bearing supports, 234
End mill, 19–20, 20f
Engraving, 20, 20f
  filled-shallow, 197
  machines, 5
  with stick fonts, 148f
  stock, 205, 206f
Enhanced machine controller, Version 2 (EMC2), 88-89
EPO. *See* Emergency power off
Epoxy-filled engraving, 198f
ER25 collet system, 10f
EZ-Router, 164, 164f

F

Fat mat, 12
FDA. *See* Food and Drug Administration
Feed rate, 123
FenderStratocaster, 207
Filled-shallow engraving, 197
Flanges, 48, 234
Food and Drug Administration, 46
FreeMill, 156
Freeware, 139
Fret wire bending tool, 193f
Furniture, 189, 190f

G

G00 (rapid linear motion), 93–95, 94f
  improper use of, 94f
G01 (linear motion at feed rate), 95–96, 95f
G02 (arc at feed rate), 96–99, 96f
G03 (arc at feed rate), 96–99, 96f
G04 (dwell), 99
G10 command (coordinate system data tool and work offset tables), 99–100
G12 (clockwise/counterclockwise circular pocket), 100
G13 (clockwise/counterclockwise circular pocket), 100
G15 (exit and enter polar mode), 100
G16 (exit and enter polar mode), 100
G17 (plane selection), 100, 100f, 101f
G18 (plane selection), 100, 100f, 101f
G19 (plane selection), 100, 100f, 101f
G20 (length units), 101
G21 (length units), 101
G28 command (return to home), 101, 102f
G28.1 (reference axes), 101–102
G30 (return to home), 101
G31 (straight probe), 102–103
G40 (cutter radius compensation), 103–104
G41 (cutter radius compensation), 103–104
G42 (cutter radius compensation), 103–104
G43 (tool length offsets), 104
G44 (tool length offsets), 104
G49 (tool length offsets), 104
G50 (scale factors), 104
G51 (scale factors), 104
G52 (temporary coordinate system offset), 105
G53 (in absolute coordinates), 105
G54 to G59 (select work offset coordinate system), 105
G59 P (select work offset coordinate system), 105
G61 (path control mode), 106
G64 (path control mode), 106
G68 (coordinate system rotation), 106
G69 (coordinate system rotation), 106
G73 canned cycle (high-speed peck drill), 106–107, 107f
G80 (modal motion cancellation), 107
G81 canned cycle drilling, 109, 110f
G81 to G89 (canned cycles), 108–109, 110f
G82 cycle (for drilling), 110
G83 (canned cycle peck drilling), 107f, 110, 111f
G85 (for boring/reaming canned cycle), 111, 111f
G86 cycle (for boring), 112

G88 cycle (for boring), 112
G89 cycle (for boring), 112
G90 (absolute distance mode), 112–113, 113f
G91 (incremental distance mode), 112–113, 113f, 114
G92, G92.1, G92.2, and G92.3 – G92 (offsets), 113–114
G93 (path control mode set), 114–115
G94 (path control mode set), 114–115
G95 (path control mode set), 114–115
G98 (canned cycle return level), 115–116
G99 (canned cycle return level), 115–116
Gantry end caps, 177, 177f
Gas metal arc welding, 19
Gas tungsten arc welding, 19
G-code commands, 10
G-code controlled system, 162
G-code editors, 90
G-code file processing, 16
G-code programming editor. *See* Programming examples in G code
Geared reducers, 46
GIMP, 140
GMAW. *See* Gas metal arc welding
Graphics programs. *See under* CAD and graphics
GTAW. *See* Gas tungsten arc welding
Guide systems. *See* Linear guide systems

H

Half-moon profile rollers, 31
Hall-effect sensor, 62
Handy touch-off tool, 82f
Hardware abstraction layer, 88
Hardware enclosure cabinets, 72f
HDU foam. *See* High-density urethane foam
HerSaf, 151
  v-bit cutters, 19
High-density urethane foam, 196, 200
Hold-down methods. *See under* CNC machines
Hole drilling, 149
Holes, 178

Homing switches, 16
Horsepower
  actual, 7
  theoretical, 6
Hybrid roller guides, 32, 32f

I

Incised carving, 150–151
Incremental distance mode (G91), 108f, 112–113, 113f
Induction motor, 6, 7
Inlay function, 149
In-rush current, 7

K

K2-CNC (company), 166

L

Lathes, 18, 166–168
Lathe topology, 131–132, 131f, 132f
Lead screw, 11, 18
  assembly, 39f
  nuts, 38–41
Lead values, 36–37, 38t
Leg-leveling system, 173, 174f
Life expectancy for precision ball screw assemblies, 244
Linear glass scales, 65
Linear guide systems, 25
  hybrid roller guides, 32, 32f
  profile rail, 27–29, 28f
  round rail, 26–27, 26f
    continuous support, 27
    from Danaher Motion, 26f
    end mounting, 27
  V-style roller, 29–32
Linear motion at feed rate (G01), 95–96, 95f
Linux operating system, 89
Load locking spring, 234
Lube holes, 234
Lubrication, 28
Luthier shops, 206

M

M codes, 116, 116t
  user-defined, 122

Index

M0, M1, M2, and M30 (program stopping and ending), 116–117
M00 and M01 commands, 99
M3 (spindle control), 117
M4 (spindle control), 117
M47 (rerun from first line), 118
M48 (override control), 118
M49 (override control), 118
M5 (spindle control), 117
M6 (tool change), 117–118, 118f
M7 (coolant control), 118
M8 (coolant control), 118
M9 (coolant control), 118
M98 (subroutine call), 118–119
M99 (return from subroutine), 119
M998 (tool change position), 121
Mach3 control software, 13, 87–88, 163, 164, 166
Machine service life, 237
Machine torch, 184
Magnetic switches, 17, 17f, 82, 82f
MasterCAM, 156
MDA precision, 168
MDF. *See* Medium density fiberboard
Mechanical probing, 13
Medium density fiberboard, 18, 197
Metal inert gas, 19
Micro switches, 16, 16f, 81–83, 81f
Microsoft Windows, 89
Microstepping drives, 75, 76f
MIG. *See* Metal inert gas
Mills, 17–18, 166–168
Mill topology, 130–131, 131f
Modal groups, 122, 122t
Monument sign, 189, 190f
Motor(s)
   encoders, 64-65, 65f
   power, 163
   servo motors, 61–63
     advantages of, 64
   stepper motors, 57–61, 64f
     advantages of, 63
     specifications for, 60t
Motor-assembly pivot, 177f
Mounted reduction unit, 178
Mounting blocks, 27
MPG from Nidec Nemicon, 81f
Multistage reduction unit, 53

N

National Electrical Manufacturers Association, 57
NEMA (National Electrical Manufacturers Association), 57
NEMA 23 motors, 58f, 178
NEMA 34 motors, 58f
NEMA/34-stepper motor, 49, 50f
NEMA motor mounting templates
   NEMA 17 footprint, 249f
   NEMA 23 footprint, 250f
   NEMA 34 footprint, 250f
Nested vector artwork, 217f
Nesting, 154–155
Nidec Nemicon Corporation, 81, 81f
Nut
   lead screw and, 38–41
   loading, 241
   mounting, 37, 241
   rotating, 42

O

Offset spur gear reduction, 46–47, 47f
Online calculator, 49, 183
Optical probing, 13
Override control, 118

P

Pacific bearing, 32
Part tabbing, 154
Particle image velocimetry, 62–63
PD. *See* Pitch diameter
Peck drilling, 107f, 110
Pendant. *See under* Controller hardware
Permissible compression loading, 238
Permissible rotational speed, 238
PID. *See* Proportional integral derivative
Pinion gear, 11, 43, 44, 45
Pinion transmission system, 42, 46
Pitch, 42–43, 47
   diameter, 43, 44, 45
PIV (particle image velocimetry), 62–63
Pixelation, 137, 137f
Planetary gear reduction, 46
Plasma cutters, 14–15
Plasma table, 164–166, 165f, 166f

Plate-bearing hole, 183
Port, definition of, 72
Power supply, 76–78, 77f, 78f
Precision rolled ball screws product
    overview, 234
  acceptable speed, 243
  accuracy, 239
  ball screw assembly selection, 235
    example, 236
  bearing support reference drawings
    (end fixity), 242
  compression load, 246
  design formulas, 237–239
  length .
    for designated ball screws, 246
    for screws, 243
  life expectancy
    for precision ball screw assemblies, 244
    for precision plus preloaded ball screw assemblies, 245
  lubrication guidelines, 241
  preload types, 240
Pressure angle (PA), 42
Probes, 13
Profile rail, 27–29, 28f
Profiling, 146–147, 145f, 146f, 153–154
Programming examples, 219–223
  in G code, 227–230
Proportional integral derivative (PID), 62
Pulley blocks, 198–199, 199f, 200f
Pulley reducers, 49–53

Q

QCAD, 139–140

R

Rack, 11
  and pinion combination, 42–46
Rack transmission system, 42, 46
Radial ball bearing, 36
Radius format command, 97
Rail assembly, 31
Raised/recessed-lettering effects, 196, 196f
Ramped entry cut, 154

Rapid linear motion (G00), 93–95, 94f
  improper use of, 94f
Ready-made CN system, 161–164
  do-it-yourself (DIY), 168
  lathes, 166–168
  mills, 166–168
  plasma table, 164–166, 165f
  router, 164–166
  vendor listing, 168–169
Reducer mechanism, 11
Reducers, 44. *See also under*
  Transmission systems
Relay cover removal job, 216–218, 218f, 220f
Relay holding tray, 218f
Remote pendant controller from Texas Micro, 80f
Rest machining, 146
Rhino, 140
Rotary A axis, 13–14
Rotary encoder, 65f
Rotary topology, 131–132, 131f, 132f
Rotating nut, 42
Round rail. *See under* Guide systems
Router, 5, 164–166
  head, 6–8, 7f
RTA motion control systems, 75

S

Screw, 11
  lead, 37–38
  lead value of, 42
  and nut. *See under* Transmission systems
Select tool, 123
Self-reversing tapping cycles, 119–121, 120t–121t
Serial-controlled spindle-speed controller card, 80f
Servo motors, 61–63
  advantages of, 64
  closed-loop functionality, 62
  tuning parameters
    PID (proportional integral derivative), 62
    PIV (particle image velocimetry), 62–63
  velocity versus acceleration, 62f

Shareware, 139
SHCS. *See* Socket head cap screws
SheetCAM (by Stable Design), 156
Shop cabinetry, 189f
Shuttle Pro, 80
Signage industry, 12
Single-pole double-throw, 16
Six-wire motors, 58
Skip-lead preload, 240
Socket head cap screws, 178
Solid-body electric guitar, 206–214, 210f, 211f, 215f
    channel with installed truss rod, 211f
    3D contouring cuts, 209f
    3D routing, 213f
    face shot of, 208f
    inlay markers, 214f
    neck headstock, 213f
    neck holding jig, 210f
    partially cut fret slots, 213f
    routing fret board, 212f
SolidWorks, 14
Sound logic, card by, 80f
SPDT. *See* Single-pole double-throw
Spindle chuck, 21f
Spindle control, 117
Spindle drive, 8
Spindle head, 8–11
    3 horsepower, 9f
Spindle speed, 123
    controller, 79, 80f
Spoil board, 131
Stand-alone regulated power supply units, 76
Stand-off supports, 27
Start-up current, 7
Stepper motors, 57–61, 64f
    advantages of, 63
    maximum performance, 61
    power supply, 60–61
    specifications for, 60t
Stick font, 148, 148f
    engraving, 206f
    text, 136, 136f
Straight-probe command, 102
Stud-mounted lettering and logos, 214–216, 215f, 216f
Subroutines, 228–230
Swarf removal, 28

T

T track grid work, 12
Table assembly, 184
Tangential ball return, 234
Teeth per inch Acme screw, 41
Temporary coordinate system offset (G52), 105
Tension loads, 242
Texas Micro circuits, 80
THC. *See* Torch height controller
Thin-wall gantry tubing, 176, 176f
Threading, 18
3D carving, 150–151
3D cutting, 138
3D routing, 202–204
TIG. *See* Tungsten inert gas
Tiger Tek CNC, 166
Timing belt, 38, 47–49, 48f
Timing pulleys, 47–49, 48f
Tool length offsets, 104
Tooling, 19–21, 21–22
Topology
    lathe, 131–132, 131f, 132f
    mill, 130–131, 131f
    rotary, 131–132, 131f, 132f
Torch height controller, 14, 15f
Tormach CNC mill, 167f
Tormach tooling system, 21, 22f, 169
Torque, 238
Torroidal-style transformer, 78f
Transmission systems
    ball screws, 41–42
    lead screw and nut, 38–41
    pulley-reduction unit construction, 49–52
        variations on, 52–53
    rack and pinion combination, 42–46, 43f
    reducers
        geared reducers, 46
        offset spur gear reduction, 46–47
        planetary gear reduction, 46
    rotating nut, 42
    screw and nut
        driving method, 37

Transmission systems *(continued)*
   end fixity, 35–36
      lead values, 36–37, 38t
      nut mounting, 37
      screw lead, 37–38
   timing belt, 47–49
   timing pulleys, 47–49
Tray socket
   cut relay in, 219f
   uncut relay in, 219f
Trucks, 28
TrueType font, 144, 150, 197
   engraving, 206f
TrueType text, 136, 136f
TTS. *See* Tormach tooling system
Tungsten inert gas, 19
TurboCAD, 139
Twelve-volt relay, 218f

U

Ubuntu, 89
Uncut relay in tray socket, 219f
Unipolar motors, 58
United States Department of Agriculture, 46
Universal serial bus technology, 162
USB technology. *See* Universal serial bus technology
USB-based microscope camera, 13
USDA. *See* United States Department of Agriculture
User-defined M codes, 122
User input, 164

V

V bit engraving, 150f
V carving, 194–195, 195f
V router bits, 194
V style cutter, 150
V style roller, 29–32
Vacuum hold-down method, 12
Variable-frequency drive, 8, 9f
Vector artwork, 215f
   nested, 217f
Vector files, 137
Vector quad, 30
Vectric software, 155
VFD. *See* Variable-frequency drive
VisualMill, 156

W

WAGO connectors, 83f
Wall support blocks, 27
Warp 9 Computers, 164, 168
Wattage, 6
Whip, 36
   critical speed, 36
Wiper kit, 234
Wiring, 83
Wood composites, 18
Wood grain pattern in MDF, 197
Woodworking router head, 7f

X

X-axis rails, 176

Y

Y-axis rail, 176

Z

Z-axis assembly, 181f, 182f
Z-axis back plate
   mounting structure of, 179
   with spacer and guide, 179
Z retraction level (for canned cycle), 115f
Z Touch-Off Tool, 83